Advanced Courses in Mathematics
CRM Barcelona

Centre de Recerca Matemàtica

Managing Editor:
Manuel Castellet

Guido Mislin
Alain Valette

Proper Group Actions and the Baum–Connes Conjecture

Springer Basel AG

Authors' addresses:

Guido Mislin
Department of Mathematics
ETH Zentrum
8092 Zürich
Switzerland

e-mail: mislin@math.ethz.ch

Alain Valette
Institut de Mathématiques
Université de Neuchâtel
Rue Emile Argand 11
2007 Neuchâtel
Switzerland
e-mail: alain.valette@unine.ch

2000 Mathematical Subject Classification 19K35, 22D25, 46L80, 55N20

A CIP catalogue record for this book is available from the
Library of Congress, Washington D.C., USA

Bibliografische Information Der Deutschen Bibliothek
Die Deutsche Bibliothek verzeichnet diese Publikation in der Deutschen Nationalbibliografie; detaillierte bibliografische Daten sind im Internet über <http://dnb.ddb.de> abrufbar.

ISBN 978-3-7643-0408-9 ISBN 978-3-0348-8089-3 (eBook)
DOI 10.1007/ 978-3-0348-8089-3

Ursprünglich erschienen bei Birkhäuser Verlag, Basel - Boston - Berlin 2003
Cover design: Micha Lotrovsky, 4106 Therwil, Switzerland
Printed on acid-free paper produced from chlorine-free pulp. TCF ∞

ISBN 978-3-7643-0408-9

9 8 7 6 5 4 3 2 1

www.birkhauser-science.com

Contents

Foreword

The Baum–Connes Conjecture for a group G predicts that a certain index map of analytical nature sets up a natural isomorphism between the G-equivariant K-homology of the universal space $\underline{E}G$ for proper G-actions and the K-theory of the reduced C^*-algebra of the group G. This conjecture has been verified for many classes of groups and is a major target of current research. Its truth implies the validity of a surprisingly large number of long-standing conjectures in group theory, functional analysis, geometry, and topology. For example, it answers affirmatively the Idempotent Conjecture, stating that if G is torsion free, then the complex group algebra $\mathbb{C}[G]$ has no idempotent other than 0 and 1, and the Novikov Conjecture, stating that the higher signatures of closed manifolds are oriented homotopy invariants.

The beauty and multidisciplinarity of this topic motivated an advanced course at the CRM from September 18 to 22, 2001. The topological and algebraic aspects of the Baum–Connes Conjecture were presented by Mislin, while the analytical aspects were presented by Valette. This book contains a revised version of our lecture notes, with emphasis on equivariant homology theories and the analytical assembly map, respectively.

Besides our indebtedness to the Centre de Recerca Matemàtica, thanks are due to Carles Casacuberta, the course coordinator, for making it possible, and to the CRM Secretaries, Consol Roca and Maria Julià, for their assistance. Special thanks go to all the participants of the course for their interest in the event and for their very helpful contributions during the afternoon sessions.

Guido Mislin
Alain Valette

This Advanced Course on Proper Group Actions was supported by the European Commission under contract number HPCF-CT-2000-00140 of the Human Potential Programme, by the Dirección General de Investigación of the Spanish Ministry for Science and Technology (PGC2000-2199-E), and by the Direcció General de Recerca of the Catalan Government (2001ARCS 00154).

Equivariant K-Homology of the Classifying Space for Proper Actions

Guido Mislin

Abstract: These notes are a compendium to a series of lectures concerning the topological aspects of the Baum–Connes Conjecture – *the left hand side of the equation* $K_*^G(\underline{E}G) \cong K_*^{top}(C_r^*(G))$ – the equivariant K-homology of $\underline{E}G$. Besides of a presentation of the material needed to compute $K_*^G(\underline{E}G)$, the reader will find an extensive discussion of many conjectures related to the Baum–Connes Conjecture.

1. Introduction and Acknowledgements

The Baum–Connes Conjecture predicts that the K-theory of the reduced C^*-algebra $C_r^*(G)$ of a group G can be computed as the K^G-homology of the so-called *universal space for proper* G*-actions*, $\underline{E}G$. Using the skeleton filtration of the G-CW-complex $\underline{E}G$, one can determine $K_*^G(\underline{E}G)$ inductively. The terms entering into the computation only involve, in a first approximation, the complex representation rings $R_{\mathbb{C}}(H)$ of the finite subgroups $H < G$. Indeed, the K^G-homology of the i-skeleton modulo the $(i-1)$-skeleton of $\underline{E}G$ has the form $\bigoplus R_{\mathbb{C}}(H)$, the sum being taken over the orbits of i-dimensional G-cells of $\underline{E}G$, and H standing for a typical isotropy group. The skeleton filtration of $\underline{E}G$ provides a spectral sequence with E^2-term certain Bredon homology groups $H_i^{\mathfrak{Fin}}(G; K_j^G(\,?\,))$, where $K_i^G(\,?\,)$ is viewed as a functor on the orbit category $\mathfrak{O}_{\mathfrak{Fin}}(G)$ of G with objects the orbits G/H, $H < G$ finite. The necessary homological algebra entering in the definition of the Bredon homology groups is presented in Section 3. This computation of the groups $K_*^G(\underline{E}G)$ via Bredon homology groups leads, when tensored with $\mathbb{Q}$, to a description of the rational K^G-groups in terms of a Chern character map (cf. Section 6). The Chern character reduces the calculation of $K_*^{top}(C_r^*(G)) \otimes \mathbb{Q}$ to a problem in homological algebra, provided G satisfies the Baum–Connes Conjecture.

Many technical results concerning the Baum–Connes Conjecture require the (discrete) groups in question to be *countable.* But, if we write G as the union of its countable subgroups $\{G_\alpha\}$, one finds for the topologically defined equivariant K-homology groups (cf. Corollary 5.7) a natural isomorphism

$$K_*^G(\underline{E}G) \cong \operatorname{colim}_\alpha K_*^{G_\alpha}(\underline{E}G_\alpha)$$

and, on the other hand, for the K-theory of the associated reduced C^*-algebras (cf. Lemma 5.3), an isomorphism

$$K_*^{top}(C_r^*(G)) \cong \operatorname{colim}_\alpha K_*^{top}(C_r^*(G_\alpha)).$$

As a result, the Baum–Connes Conjecture (cf. Conjecture 5.9) holds for a group G if it holds for all of its countable subgroups. Thus, it is possible to drop the annoying countability assumption in many places.

It is amazing how many basic conjectures in algebra and topology are related to the Baum–Connes Conjecture, some very directly, others in a more indirect way. The reader will find a scheme at the end of the Appendix (Section 7) illustrating these connections in the form of a connected graph. We will discuss in particular the following conjectures:

Idempotent Conjecture **IC**,
Strong Idempotent Conjecture **SIC**,
Trace Conjecture **TC**,
Strong Trace Conjecture **STC**,
Novikov Conjecture **NC**,
Strong Novikov Conjecture **SNC**,
Gromov–Lawson–Rosenberg Conjecture **GLRC**,
Zero-In-The-Spectrum Conjecture **0∊SC**,
Strong Zero-In-The-Spectrum Conjecture **S0∊SC**,
Zero Divisor Conjecture **ZDC**,
Atiyah Conjecture **AC**,
Strong Atiyah Conjecture **SAC**,
Embedding Conjecture **EC**,
Bass Trace Conjecture **BTC**,
Weak Bass Trace Conjecture **WBTC**,
Strong Bass Trace Conjecture **SBTC** and the
Projective Class Group Conjecture **PCGC**.

Many thanks go to the people who pointed out mistakes in earlier versions of these notes and suggested improvements, notably Michel Matthey and Wolfgang Lück, whose help was very generous. Thanks go to all the people, who followed patiently the presentation at the Summer School at the CRM of Barcelona in September 2001. Particular thanks go to Indira Chatterji, whose help in preparing these notes was crucial and whose active work at the seminars in the course of the preparations in Zurich helped tremendously. Finally, a special *Thank You* to Alain Valette, for his great patience in explaining to me the intricate points of the analytic version of the assembly map and the basic facts on group C^*-algebras.

2. Odds and Ends on G-Spaces

In the sequel we will only be dealing with actions of *discrete* groups. Most of the definitions and results admit obvious extensions to the case of topological group actions (cf. tom Dieck's book [35]).

Let G be a discrete topological group. A G-space is a topological space X equipped with a (continuous) left G-action

$$G \times X \longrightarrow X, \quad (g, x) \mapsto gx$$

satisfying $ex = x$ and $g(hx) = (gh)x$.

The *stabilizer* $G_x < G$ of a point $x \in X$ is the subgroup $\{g \in G \mid gx = x\}$. The cartesian product $X \times Y$ of two G-spaces is a G-space via the diagonal G-action $(g, (x, y)) \mapsto (gx, gy)$; note that $G_{(x,y)} = G_x \cap G_y$. The mapping space $\mathrm{map}(X, Y)$ of (continuous) maps $X \to Y$, endowed with the compact-open topology, is a G-space via the diagonal action $(g, \phi) \mapsto g\phi(g^{-1}x)$. We recall that for Y locally compact (e.g. Y=G), the functor $X \mapsto X \times Y$ is left adjoint to $Z \mapsto \mathrm{map}(Y, Z)$ in the category of all topological spaces.

If $H < G$ is a subgroup, we write X^H for the subspace of H-fixed points of X, and X/H for the space of H-orbits $\{Hx \mid x \in X\}$ (with the quotient topology). If $N(H)$ denotes the normalizer $\{g \in G \mid gH = Hg\}$ of H in G, then the G-action on X restricts to an $N(H)$-action on X^H, with H acting trivially. Thus X^H is an $N(H)/H$-space; the group $N(H)/H$ is called the Weyl group of $H < G$, and it is usually denoted by $W(H)$.

The space G/H of left cosets of G is a G-space via $(g, kH) \mapsto gkH$. Every discrete G-space is a disjoint union of such G-spaces. For $K < G$, the fixed point space $(G/H)^K$ consists of all cosets gH satisfying $KgH = gH$, i.e. $g^{-1}Kg < H$.

The following two constructions are basic. Let X be an H-space, $H < G$. The *induced* G-space is

$$\mathrm{Ind}_H^G X = G \times_H X$$

which is the space of orbits $(G \times X)/H$ where H acts on $G \times X$ via $(h, (g, x)) \mapsto (gh^{-1}, hx)$. The left G-action on $G \times X$ then passes to a G-action on $G \times_H X$: $(g, \overline{(k, x)}) \mapsto \overline{(gk, x)}$. If Z is any G-space, then the H-maps $X \to Z$ correspond bijectively to G-maps $G \times_H X \to Z$ (so that $X \mapsto G \times_H X$ is left adjoint to the forgetful functor from G-spaces to H-spaces). For instance, if $K < H < G$, the obvious H-map $H/K \to G/K$ extends to a bijection of G-spaces $G \times_H (H/K) = G/K$. Dual to induction, we can associate with an H-space X the *coinduced* G-space of H-maps

$$\mathrm{map}_H(G, X)$$

where G is considered as an H-space via the left H-action and the G-action on $\mathrm{map}_H(G, X)$ is given on $\phi \in \mathrm{map}_H(G, X)$ by $g\phi : G \to X, y \mapsto \phi(yg)$ so that $(g\phi)(hy) = \phi(hyg) = h\phi(yg) = h((g\phi)(y))$, i.e. $g\phi$ is indeed an H-map. For Z any G-space, the H-maps $Z \to X$ correspond bijectively to G-maps $Z \to \mathrm{map}_H(G, X)$, making coinduction right adjoint to the forgetful functor from G-spaces to H-spaces. In case X is already a G-space, which one can consider as an H-space by restricting the action to H, induction and coinduction take the following simple form:

$$G \times_H X \xrightarrow{\cong} (G/H) \times X,$$

the map being obtained by extending the H-map $x \mapsto (H, x)$ to a G-map. Dually, one has for a G-space X a natural bijection

$$\mathrm{map}(G/H, X) \xrightarrow{\cong} \mathrm{map}_H(G, X)$$

obtained by coextending the H-map $\mathrm{map}\,(G/H, X) \to X, \phi \mapsto \phi(H)$ to a G-map.

We will later make use of a slightly more general version of induction, which is defined as follows. Let $\alpha : H \to G$ be an arbitrary homomorphism and let X be an H-space. Then the induced G-space

$$\mathrm{Ind}_\alpha X = G \times_\alpha X$$

is the G-space whose underlying space is $(G \times X)/H$, where H acts on the product by $h(g, x) = (g\alpha(h^{-1}), hx)$ and where the G-action on $\mathrm{Ind}_\alpha X$ is given by the left G-action on $G \times X$; similarly, one can define the coinduced space $\mathrm{Coind}_\alpha X$.

Another important construction which we will use is the *join* of G-spaces. We will just state some basic facts. Details can be found in [35, Chap. I, Sec. 6]. For each $n \geq 0$ let σ^n denote the standard n-simplex in $\mathbb{R}^{n+1}$:

$$\sigma^n = \left\{ (t_0, \ldots, t_n) \in \mathbb{R}^{n+1} \,\middle|\, \sum t_i = 1,\ t_i \geq 0 \right\}.$$

Given spaces $X_0, \ldots, X_n$, the join $X_0 * \cdots * X_n$ is defined to be the identification space

$$(\sigma^n \times X_0 \times \cdots \times X_n)/\sim$$

where

$$(t_0, \ldots, t_n; x_0, \ldots, x_n) \sim (t'_0, \ldots, t'_n; x'_0, \ldots, x'_n)$$

if and only if for each i, either $(t_i, x_i) = (t'_i, x'_i)$ or $t_i = t'_i = 0$. For instance, $X * \{pt\}$ is just the cone CX on X. If the X_i are G-spaces then the join inherits a G-action by letting G act diagonally (trivially on the σ^n). For spheres one has $S^n * S^m \cong S^{n+m+1}$.

A G-space X is called *free*, if X^H is empty for every $H < G, H \neq \{e\}$. To avoid bad free G-spaces (like $\mathbb{Q}$ acting on $\mathbb{R}$ by translation) and to be able to extend some results to certain non-free actions, we will make use of a notion of *proper* G-spaces. In the literature, one can find a variety of definitions of proper G-spaces, many of which agree under various topological restrictions. We will call a G-space proper, if its G-action is locally induced from the action of a finite subgroup in the following sense.

Definition 2.1. A G-space is called *proper*, if there are *finite* subgroups $H_\iota < G$ and open H_ι-invariant subspaces $X_\iota \subset X$ such that the natural G-maps $G \times_{H_\iota} X_\iota \to X$ are G-homeomorphisms onto their images, and

$$X = \bigcup_\iota G \times_{H_\iota} X_\iota.$$

Clearly, all point stabilizers for a proper G-space are finite. If the G-space X is free and proper, then the projection $X \to X/G$ is a local homeomorphism (Galois covering, locally trivial principal G-bundle). Note also that for any $\alpha : H \to G$ and any proper H-space X, the induced G-space $\mathrm{Ind}_\alpha X$ is a proper G-space. (The reader who is interested in the relationship between different definitions of *proper actions* on locally compact Hausdorff spaces should consult [102] and [25].)

A G-map $f : X \to Y$ is called a *weak G-equivalence* if for every $H < G$ its restriction $f^H : X^H \to Y^H$ is a weak homotopy equivalence (i.e. $\pi_i(X^H, x_0) \to \pi_i(Y^H, f^H(x_0))$ is bijective for $i \geq 0$ and every $x_0 \in X^H$).

The notions of G-homotopy and G-homotopy equivalence are defined in an obvious way. For certain good types of G-spaces, weak G-equivalences are actually G-homotopy equivalences; this is in particular so for the class of G-CW-complexes, which we will discuss now.

A *G-CW-complex* consists of a G-space X together with a filtration $X^0 \subset X^1 \subset X^2 \subset \cdots \subset X$ by G-subspaces such that the following axioms hold:

1. Each X^n is closed in X.
2. $\bigcup_{n\in\mathbb{N}} X^n = X$.
3. X^0 is a discrete subspace of X.
4. For each $n \geq 1$ there is a discrete G-space Δ_n together with G-maps $f : S^{n-1} \times \Delta_n \to X^{n-1}$ and $\widehat{f} : B^n \times \Delta_n \to X^n$ such that the following diagram is a push-out diagram:

$$\begin{array}{ccc} S^{n-1} \times \Delta_n & \xrightarrow{f} & X^{n-1} \\ \downarrow & & \downarrow \\ B^n \times \Delta_n & \xrightarrow{\widehat{f}} & X^n. \end{array}$$

5. A subspace Y of X is closed if and only if $Y \cap X^n$ is closed for each $n \geq 0$.

Here, we write S^{n-1} and B^n for the standard unit sphere and unit ball in Euclidean n-space, and the vertical maps in the diagram are inclusions. In case of $G = \{e\}$, a G-CW-complex is just an ordinary CW-complex. The notion of a G-subcomplex of a G-CW-complex X is defined in the obvious way; similarly for G-CW-pairs. It is useful to adopt the conventions $X^{-1} = \emptyset$ and $\Delta_0 = X^0$.

A G-CW-complex X is said to be *finite dimensional* if and only if $X^n = X$ for some n, in which case the dimension of X is the least $n \geq -1$ for which this happens (the empty space has dimension -1). For the finite dimensional case, Axiom (5) is redundant. Clearly, if X is a G-CW-complex, then X/G is an ordinary CW-complex; X is called G-finite, or *cocompact*, if X/G is a finite CW-complex.

The product of two CW-complexes X and Y in the product topology is in general not a CW-complex. But if one equips $X \times Y$ with the *weak* topology ($A \subset X \times Y$ closed if and only if A intersects every $X_\alpha \times Y_\beta$ in a closed subset for all finite subcomplex X_α of X and Y_β of Y), then $X \times Y$ is a CW-complex in an obvious way: the n-cells of $X \times Y$ are the products of k-cells of X with $(n-k)$-cells of Y, $0 \leq k \leq n$. A similar remark applies to the join construction: $X * Y$ is a

CW-complex, if it is equipped with the weak topology, coming from the joins of finite subcomplexes. By abuse of notation we will still just write $X \times Y$ resp. $X * Y$ when we deal with CW-complexes, meaning that the product, resp. join, is given the weak topology.

The following equivariant version of the Whitehead Theorem is a very useful and basic result.

Theorem 2.2. (J.H.C. Whitehead, cf. [96, Chapter I]) *A G-map $f : X \to Y$ between G-CW-complexes is a G-homotopy equivalence if for all $H < G$ and all $x_0 \in X^H$ the induced map $\pi_*(X^H; x_0) \to \pi_*(Y^H; f(x_0))$ is bijective (i.e., if f is a weak G-equivalence).*

Remark 2.3. If one is only interested in G-homotopy types up to weak equivalence, it is good to know that the usual geometric realization functor from spaces to CW-complexes extends to a functor Γ from G-spaces to G-CW-complexes, together with a natural transformation $\Gamma(X) \to X$, which is a weak equivalence for every G-space X (cf. [96]). This can be used, for instance, to turn mapping spaces $\mathrm{map}(X, Y)$ of G-CW-complexes into G-CW-complexes $\Gamma(\mathrm{map}(X, Y))$ with homotopy groups as one expects. We will, for instance, use this device to consider loop spaces of G-CW-complexes as G-CW-complexes.

A G-CW-complex is a proper G-space, if and only if all point stabilizers are finite. The following construction of the standard model of a *universal* proper G-CW-complex $\underline{E}G$ is fundamental (cf. tom Dieck [34] and [35]). Let M be the zero-dimensional G-CW-complex given by the disjoint union of all cosets G/H, H finite. Let $M(n)$ denote the n-fold join of M, with the CW-topology; it is an $(n-1)$-dimensional proper G-CW-complex. There are obvious inclusions $M(n) \to M(n+1)$ and we put

$$\underline{E}G := \bigcup_{n \in \mathbb{N}} M(n),$$

where the space is considered as a G-CW-complex by giving it the weak topology ($A \subset \underline{E}G$ is closed if and only if $A \cap M(n)$ is closed for all n). If $H < G$ is finite, then $(\underline{E}G)^H$ is the union of subspaces $M(n)^H$, which are $(n-2)$-connected, being the join of n non-empty CW-complexes M^H. Thus $(\underline{E}G)^H$ is contractible (every map $S^j \to (\underline{E}G)^H$ has its image in some $M(n)^H$ because S^j is compact; but $M(n)^H$ is j-connected for $n \geq j+2$). Our construction of $\underline{E}G$ is natural with respect to group homomorphisms $\phi : G \to H$; for $K < G$ finite, ϕ induces a G-map $G/K \to H/\phi(K)$, where $H/\phi(K)$ is considered as a G-space via $(g, h\phi(K)) \mapsto \phi(g)h\phi(K)$. By passing to a colimit of joins of cosets, one obtains a G-map

$$\underline{E}\phi : \underline{E}G \longrightarrow \underline{E}H.$$

It is an easy consequence of the equivariant Whitehead Theorem 2.2 that $\underline{E}G$ enjoys the following *universal property*, which characterizes it up to G-homotopy.

Theorem 2.4. *Let X be a proper G-CW-complex. Then, up to G-homotopy, there is a unique G-map $X \to \underline{E}G$.*

Proof. One can construct a G-map $X \to \underline{E}G$ by first observing that the projection $\mathrm{pr}_X : X \times \underline{E}G \to X$ is a G-homotopy equivalence by 2.2, since $(\underline{E}H)^H$ is contractible for every finite $H < G$. There is thus a G-homotopy class $\mathrm{pr}_{\underline{E}G} \circ \mathrm{pr}_X^{-1} : X \to X \times \underline{E}G \to \underline{E}G$. For the uniqueness part we first observe that the diagonal map $\Delta : \underline{E}G \to \underline{E}G \times \underline{E}G$ is a G-homotopy equivalence by 2.2. Therefore, the two projections $\mathrm{pr}_i : \underline{E}G \times \underline{E}G \to \underline{E}G$, $i = 1, 2$ are homotopic. If $\alpha, \beta : X \to \underline{E}G$ are two G-maps, then the diagram

$$X \xrightarrow{\{\alpha,\beta\}} \underline{E}G \times \underline{E}G \xrightarrow{\mathrm{pr}_i} \underline{E}G, \quad i = 1, 2$$

shows that $\alpha = \mathrm{pr}_1 \circ \{\alpha, \beta\}$ is G-homotopic to $\beta = \mathrm{pr}_2 \circ \{\alpha, \beta\}$. □

Another way of proving Theorem 2.4 would be by applying equivariant obstruction theory (see 3.11).

One should think of Theorem 2.4 as expressing that in the homotopy world of proper G-CW-complexes, $\underline{E}G$ plays the same role as does a point in the homotopy world of ordinary CW-complexes. We will use the notation "$\underline{E}G$" to denote any G-CW-complex G-homotopy equivalent to the functorial construction of $\underline{E}G$ which we just described. For instance, $\{pt\}$ is an $\underline{E}G$ for a finite group G, and for F a free group one has a tree model for $\underline{E}F$ (i.e. a 1-dimensional proper, contractible F-CW-complex).

Remark 2.5. A similar construction provides a universal G-space $E\mathfrak{F}$ for an arbitrary non-empty family $\mathfrak{F}$ of subgroups of G which is closed under conjugation and passing to subgroups. One replaces in the construction above the M by $\coprod_{K\in\mathfrak{F}} G/K$ and obtains a G-CW-complex $E\mathfrak{F}$ such that $(E\mathfrak{F})^K$ is contractible for $K \in \mathfrak{F}$ and empty, if $K < G$ does not belong to $\mathfrak{F}$. It is universal in the sense of Theorem 2.4 for G-CW-complexes X having stabilizers in $\mathfrak{F}$: every such X admits a G-map $X \to E\mathfrak{F}$, unique up to G-homotopy.

3. Bredon Cohomology

The definition of Bredon cohomology groups for G-CW-complexes in case of a finite group G goes back to Bredon [16]. The definition is easily extended to the case of infinite groups and plays there an important role, because of its use in equivariant obstruction theory [96]. We recall here the basic definitions. Let $\mathfrak{F}$ be a non-empty family of subgroups of G (there are at this point no further assumptions on the family $\mathfrak{F}$). Write $\mathfrak{O}_{\mathfrak{F}}(G)$ for the orbit category, whose objects are left coset spaces G/K with $K \in \mathfrak{F}$, and morphism sets $\mathrm{mor}(G/K, G/L)$ are the G-maps $G/K \to G/L$. A G-map $\phi : G/K \to G/L$ is determined by $\phi(K) \in G/L$ and, because of $\phi(gK) = g\phi(K)$, one must have $\phi(K) \in (G/L)^K$. Conversely, a point $xL \in (G/L)^K$ determines a unique G-map $G/K \to G/L$, by $gK \to gxL$. Thus, one has a natural bijection

$$\mathrm{mor}(G/K, G/L) \to (G/L)^K, \quad f \mapsto f(K).$$

In particular, the automorphism group $\mathrm{Aut}(G/K)$ is isomorphic to $N(K)/K = W(K)$, by mapping $\phi \in \mathrm{Aut}(G/K)$ to xK, if $\phi(K) = x^{-1}K$; note that if K is finite, $\mathrm{mor}(G/K, G/K) = \mathrm{Aut}(G/K)$: all morphisms are isomorphisms in this case.

Definition 3.1. We write $\mathrm{Mod}_{\mathfrak{F}}\text{-}G$ for the category of contravariant functors $\mathfrak{O}_{\mathfrak{F}}(G) \to \mathfrak{Ab}$, with $\mathfrak{Ab}$ denoting the category of abelian groups.

We also write $\mathfrak{O}_{\mathfrak{All}}(G)$ and $\mathrm{Mod}_{\mathfrak{All}}\text{-}G$ if $\mathfrak{F}$ is the family $\mathfrak{All}$ of *all* subgroups of G.

Morphisms in $\mathrm{Mod}_{\mathfrak{F}}\text{-}G$ are natural transformations of functors. Thus, if $M, N \in \mathrm{Mod}_{\mathfrak{F}}\text{-}G$, then $\Phi : M \to N$ is given by a family of homomorphisms of abelian groups $\Phi(G/H) : M(G/H) \to N(G/H)$ such that for every $\phi \in \mathrm{mor}(G/K, G/L)$ one has a commutative diagram

$$\begin{array}{ccc} M(G/K) & \xrightarrow{\Phi(G/K)} & N(G/K) \\ {\scriptstyle M(\phi)}\uparrow & & \uparrow{\scriptstyle N(\phi)} \\ M(G/L) & \xrightarrow{\Phi(G/L)} & N(G/L). \end{array}$$

A special case to keep in mind is the case where $\mathfrak{F}$ consists of the trivial subgroup $\{e\}$ only. The category $\mathfrak{O}_{\mathfrak{F}}(G)$ has then a single object $G/\{e\} = G$, and the morphisms $\mathrm{mor}(G, G)$ correspond to group elements $x \in G$, which define G-maps $\theta_x : g \mapsto gx$. Because of the contravariant nature of $M \in \mathrm{Mod}_{\mathfrak{F}}\text{-}G$, $M(G)$ is naturally a left G-module ($x \in G$ acts via $M(\theta_x)$ on $M(G)$ so that $M(\theta_{xy}) = M(\theta_y\theta_x) = M(\theta_x)M(\theta_y)$). Thus, the category $\mathrm{Mod}_{\mathfrak{F}}\text{-}G$ is naturally equivalent to the category G-Mod of left G-modules.

Returning to the general situation, we note that an abelian group A defines a *constant* functor $\underline{A} \in \mathrm{Mod}_{\mathfrak{F}}\text{-}G$: $\underline{A}(G/H) = A$ for every $H \in \mathfrak{F}$ and its value on morphisms is the identity homomorphism of A. It obviously has the property that for any $M \in \mathrm{Mod}_{\mathfrak{F}}\text{-}G$

$$\mathrm{mor}(\underline{A}, M) \cong \mathrm{Hom}(A, \lim_{\mathfrak{O}_{\mathfrak{F}}^{op}(G)} M(G/H)),$$

where the (inverse) limit is taken over the opposite category $\mathfrak{O}_{\mathfrak{F}}^{op}(G)$ of $\mathfrak{O}_{\mathfrak{F}}(G)$. We can express this by saying that the evaluation functor

$$\epsilon : \mathrm{Mod}_{\mathfrak{F}}\text{-}G \longrightarrow \mathfrak{Ab}, \quad M \mapsto \lim_{\mathfrak{O}_{\mathfrak{F}}^{op}(G)} M(G/H)$$

is right adjoint to the embedding

$$\mathfrak{Ab} \longrightarrow \mathrm{Mod}_{\mathfrak{F}}\text{-}G, \quad A \mapsto \underline{A}.$$

The category $\mathrm{Mod}_{\mathfrak{F}}\text{-}G$ is abelian. In particular, the sets $\mathrm{mor}(M, N)$ have a natural abelian group structure, coming from the abelian group structure on each $\mathrm{Hom}(M(G/H), N(G/H))$. The kernel of a morphism $M \to N$ is given by the functor, whose value on G/K is $\mathrm{Ker}(M(G/K) \to N(G/K))$; similarly for cokernels. Thus, a sequence $M \to N \to L$ is exact in $\mathrm{Mod}_{\mathfrak{F}}\text{-}G$, if and only if for every $K \in \mathfrak{F}$ the sequence $M(G/K) \to N(G/K) \to L(G/K)$ is exact. One can define chain complexes, resolutions, chain homotopies etc. in the usual manner (it is assumed that the reader has some basic knowledge of homological algebra in categories of modules). An object $P \in \text{Mod}_{\mathfrak{F}}\text{-}G$ is called *projective* if the functor

$$\text{mor}(P, ?) : \text{Mod}_{\mathfrak{F}}\text{-}G \longrightarrow \mathfrak{Ab}$$

is exact. The following construction gives rise to projective objects. Let $K \in \mathfrak{F}$ and consider the contravariant functor

$$P_K : \mathfrak{O}_{\mathfrak{F}}(G) \longrightarrow \mathfrak{Ab}, \quad G/H \mapsto \mathbb{Z}[\text{mor}(G/H, G/K)].$$

Here, $\mathbb{Z}[\text{mor}(G/H, G/K)]$ denotes the free abelian group with basis $\text{mor}(G/H, G/K)$; note that a G-map $G/H \to G/L$ yields an obvious homomorphism $P_K(G/L) \to P_K(G/H)$ defining P_K on morphisms. If $f : P_K \to M$ is a morphism, we can evaluate the resulting homomorphism $\mathbb{Z}[\text{mor}(G/K, G/K)] \to M(G/K)$ at the identity $1 \in \text{mor}(G/K, G/K)$ to obtain an element $f(G/K)(1) := \text{ev}_K(f) \in M(G/K)$; we claim that this evaluation map

$$\text{ev}_K : \text{mor}(P_K, M) \to M(G/K)$$

is bijective. To see this it is useful to think of $\text{mor}(G/K, G/K)$ as a *monoid* (composition of morphisms gives an associative binary operation with identity; if, for instance, K is finite, this monoid is a group, naturally isomorphic to the Weyl group $W(K)$). The abelian group $M(G/K)$ is a module over the monoid ring $\mathbb{Z}[\text{mor}(G/K, G/K)]$ and therefore every $x \in M(G/K)$ gives rise to a unique $\phi(x) : P_K(G/K) \to M(G/K)$ mapping $1 \in \text{mor}(G/K, G/K)$ to x, because $\phi(x)$ has to be a module homomorphism over the monoid ring $P_K(G/K)$. Now $\phi(x)$ extends to a morphism $f : P_K \to M$ as follows. We define $f(G/H)$ on basis elements $\lambda \in \text{mor}(G/H, G/K) \subset P_K(G/H)$ by noting that $\lambda = P_K(\lambda)(1)$ so that we can put $f(G/H)(\lambda) = M(\lambda)(\phi(x)(1))$:

$$\begin{array}{ccc} \lambda \in P_K(G/H) & \xrightarrow{f(G/H)} & M(G/H) \\ {\scriptstyle P_K(\lambda)}\uparrow & & {\scriptstyle M(\lambda)}\uparrow \\ 1 \in P_K(G/K) & \xrightarrow{\phi(x)} & M(G/K). \end{array}$$

It is now quite obvious that ev_K is a bijection. As a result, $\text{mor}(P_K, ?)$ maps a short exact sequence $M \rightarrowtail N \twoheadrightarrow L$ to a short exact sequence $M(G/K) \rightarrowtail N(G/K) \twoheadrightarrow L(G/K)$, showing that P_K is indeed projective. Next, we want to prove that $\text{Mod}_{\mathfrak{F}}\text{-}G$ has enough projectives, meaning that for every $M \in \text{Mod}_{\mathfrak{F}}\text{-}G$ there is an epimorphism $P \to M$ with P projective. Note that the sum (coproduct) of a family of objects M_α is given by $(\coprod M_\alpha)(G/H) = \bigoplus M_\alpha(G/H)$ with obvious inclusions. Given M, we first define

$$\Phi_K(G/K) : (\textstyle\coprod_{x \in M(G/K)} P_K)(G/K) \xrightarrow{\phi(x)} M(G/K)$$

by putting the component of $\Phi_K(G/K)$ corresponding to the index x in $M(G/K)$ to be the map $\phi(x)$ defined earlier, so that $\Phi_K(G/K)$ is onto. As above, $\Phi_K(G/K)$

extends to a unique morphism $\Phi_K : \coprod P_K \to M$. Now we repeat this construction for every $H \in \mathfrak{F}$ and define a morphism Φ with components Φ_H:

$$\Phi : \coprod_{H\in\mathfrak{F}} (\coprod_{M(G/H)} P_H) \longrightarrow M ,$$

which is an epimorphism of a projective object onto M.

Every $M \in \mathrm{Mod}_{\mathfrak{F}}\text{-}G$ admits therefore a projective resolution and projective resolutions are unique up to chain homotopy. We write $P_*(M) \twoheadrightarrow M$ for a projective resolution of M. For each $N \in \mathrm{Mod}_{\mathfrak{F}}\text{-}G$ this yields an ordinary cochain complex $\mathrm{mor}(P_*(M), N)$. Thus one can define derived functors $\mathrm{Ext}^i(M,N)$, which are contravariant in the first variable and covariant in the second, by choosing a projective resolution $P_*(M) \twoheadrightarrow M$ an putting

$$\mathrm{Ext}^i(M,N) = H^i(\mathrm{mor}(P_*(M),N)), \quad i \geq 0.$$

Bredon cohomology groups are now defined as follows.

Definition 3.2. The *Bredon cohomology groups of G with coefficients in $M \in \mathrm{Mod}_{\mathfrak{F}}\text{-}G$* are given by

$$H^i_{\mathfrak{F}}(G;M) = \mathrm{Ext}^i(\underline{\mathbb{Z}},M), \quad i \geq 0.$$

In case where $\mathfrak{F}$ consists of the trivial group only, this reduces to the usual cohomology groups, via the identification of $\mathrm{Mod}_{\mathfrak{F}}\text{-}G$ with the category of G-modules. On the other hand, if $\mathfrak{F}$ contains G, $\underline{\mathbb{Z}} = P_G$ is projective and thus $H^i_{\mathfrak{F}}(G;M) = 0$ for all $i > 0$ and $H^0_{\mathfrak{F}}(G;M) = M(G/G)$. In general one has

$$H^0_{\mathfrak{F}}(G;M) = \mathrm{mor}(\underline{\mathbb{Z}},M) = \lim_{G/K\in\mathfrak{O}^{op}_{\mathfrak{F}}(G)} M(G/K),$$

the (inverse) limit being taken over the opposite orbit category (one can therefore think of the Bredon cohomology groups as the right derived functors of this (inverse) limit functor). A short exact sequence $M \rightarrowtail N \twoheadrightarrow L$ gives rise to a long exact sequence of Bredon cohomology groups

$$\cdots \to H^i_{\mathfrak{F}}(G;M) \to H^i_{\mathfrak{F}}(G;N) \to H^i_{\mathfrak{F}}(G;L) \to H^{i+1}_{\mathfrak{F}}(G;M) \to \cdots$$

induced from the short exact sequence of cochain complexes

$$\mathrm{mor}(P_*(\underline{\mathbb{Z}}),M) \rightarrowtail \mathrm{mor}(P_*(\underline{\mathbb{Z}}),N) \twoheadrightarrow \mathrm{mor}(P_*(\underline{\mathbb{Z}}),L).$$

A *G-CW*-complex gives rise to projectives as follows. Let $\mathfrak{F}(X)$ denote the family of subgroups of G which occur as stabilizers of the G-action on X and let $\mathfrak{F}$ be a family of subgroups of G containing $\mathfrak{F}(X)$. The cellular chain complex $C_*(X)$ consists of G-modules such that

$$C_i(X) = \mathbb{Z}[\Delta_i],$$

where Δ_i is the G-set occurring in the definition of the i-skeleton of X. If $H < G$ then

$$C_i(X^H) = \mathbb{Z}[\Delta_i^H]$$

and, as Δ_i is a disjoint union of orbits G/K_α with $K_\alpha \in \mathfrak{F}(X) \subset \mathfrak{F}$, we see that Δ_i^H is the disjoint union of sets of the form $(G/K_\alpha)^H = \operatorname{mor}(G/H, G/K_\alpha)$. This shows that

$$C_i(X^H) \cong \mathbb{Z}[\amalg_\alpha \operatorname{mor}(G/H, G/K_\alpha)] \cong \bigoplus_\alpha P_{K_\alpha}(G/H).$$

Moreover, a G-map $\phi : G/K \to G/L$ induces a cellular map $X^L \to X^K$ by mapping $x \in X^L$ to $gx \in X^K$, where g is determined by the equation $\phi(K) = gL$ (thus $g^{-1}Kgx = x$ since $g^{-1}Kg < L$ and therefore $Kgx = gx$, i.e. $gx \in X^K$). It follows that

$$\underline{C_i(X)} : \mathfrak{O}_{\mathfrak{F}}(G) \longrightarrow \mathfrak{Ab}, \quad G/H \mapsto C_i(X^H)$$

is a contravariant functor, and since we assume $\mathfrak{F}(X) \subset \mathfrak{F}$, it is a sum of contravariant functors of the form P_{K_α} with K_α being the stabilizer of some i-cell of X. In particular, $\underline{C_i(X)}$ is projective for every i, whether it is considered as an object in $\operatorname{Mod}_{\mathfrak{F}}$-$G$ or $\operatorname{Mod}_{\mathfrak{F}(X)}$-$G$. Note also that there is a natural augmentation map $\underline{C_0(X)} \to \underline{\mathbb{Z}}$, defined on an object G/K by mapping each basis element of $C_0(X^K)$ to $1 \in \mathbb{Z}$; if X^K is empty, this is just the zero map.

Definition 3.3. Let X be a G-CW-complex and $\mathfrak{F}$ a family of subgroups of G containing the family $\mathfrak{F}(X)$ of isotropy groups of X. Let $M \in \operatorname{Mod}_{\mathfrak{F}}$-$G$. Then the *Bredon cohomology groups of X with coefficients in M* are the groups

$$H^i_{\mathfrak{F}}(X; M) = H^i(\operatorname{mor}(\underline{C_*(X)}, M)), \quad i \geq 0.$$

If $M \in \operatorname{Mod}_{\mathfrak{F}}$-$G$ then we can consider M as an object in $\operatorname{Mod}_{\mathfrak{F}(X)}$-$G$ by restriction. The chain complex $\operatorname{mor}(\underline{C_*(X)}, M)$ is the same, when viewed as a chain complex over $\operatorname{Mod}_{\mathfrak{F}(X)}$-$G$, with M replaced by its restriction. The definition of the Bredon cohomology groups is therefore independent of $\mathfrak{F}$ as long as $\mathfrak{F}$ contains $\mathfrak{F}(X)$; it is sometimes convenient in the applications to permit families $\mathfrak{F}$ larger that just $\mathfrak{F}(X)$. For instance, if X is a free G-CW-complex,

$$H^*_{\mathfrak{F}}(X; M) \cong H^*_{\{\{e\}\}}(X; M) \cong H^*\left(\operatorname{Hom}_G(C_*(X), M(G/\{e\}))\right),$$

which is the equivariant cohomology of X with coefficients in the G-module $M(G/\{e\})$=$M(G)$, usually denoted by $H^*_G(X; M(G))$. In case of the constant functor $M = \underline{\mathbb{Z}}$ this reduces to ordinary cohomology $H^*(X/G; \mathbb{Z})$. On the other extreme, if X is a CW-complex on which G acts trivially, then we can take $\mathfrak{F} = \mathfrak{All}$ to make certain that $\mathfrak{F}(X)$ is contained in $\mathfrak{F}$, and one finds

$$H^*_{\mathfrak{All}}(X; M) = H^*(X; M(G/G)),$$

the ordinary cohomology of X with coefficients in the abelian group $M(G/G)$=$M(\{*\})$. Returning to the general situation, we note that $\underline{C_*(X)}$ is a complex of projectives in the category $\operatorname{Mod}_{\mathfrak{F}}$-$G$ so that a short exact sequence $M \rightarrowtail N \twoheadrightarrow L$ in $\operatorname{Mod}_{\mathfrak{F}}$-$G$ gives rise to a short exact sequence of cochain complexes

$$\operatorname{mor}(\underline{C_*(X)}, M) \rightarrowtail \operatorname{mor}(\underline{C_*(X)}, N) \twoheadrightarrow \operatorname{mor}(\underline{C_*(X)}, L)$$

and hence to an associated long exact cohomology sequence

$$\cdots \to H^i_{\mathfrak{F}}(X;M) \to H^i_{\mathfrak{F}}(X;N) \to H^i_{\mathfrak{F}}(X;L) \to H^{i+1}_{\mathfrak{F}}(X;M) \to \cdots .$$

The connection with Bredon cohomology of groups is given by the following lemma.

Lemma 3.4. *For any group G and any non-empty family $\mathfrak{F}$ of subgroups of G which is closed under conjugation and taking subgroups, the chain complex $\underline{C_*(E\mathfrak{F})}$ defines a projective resolution of the constant functor $\underline{\mathbb{Z}} \in \mathrm{Mod}_{\mathfrak{F}}$-$G$.*

Proof. We know already that the modules $\underline{C_i(E\mathfrak{F})}$ are all projective. A sequence of modules in $\mathrm{Mod}_{\mathfrak{F}}$-$G$ is exact if and only if it is exact when evaluated at every G/H. But since for any $H \in \mathfrak{F}$ the complex $\underline{C_*(E\mathfrak{F})}(G/H) = C_*(E\mathfrak{F}^H)$ is acyclic, the claim follows. □

As a consequence, the Bredon cohomology groups of the space $E\mathfrak{F}$ are those of the group G.

Corollary 3.5. *Let $\mathfrak{F}$ be a non-empty family of subgroups of G closed under conjugation and taking subgroups. Let $E\mathfrak{F}$ be the universal G-CW-complex with isotropy groups the family $\mathfrak{F}$ and $M \in \mathrm{Mod}_{\mathfrak{F}}$-$G$. Then*

$$H^i_{\mathfrak{F}}(E\mathfrak{F};M) \cong H^i_{\mathfrak{F}}(G;M).$$

If (X, A) is a G-CW-pair, the relative chain complex $\underline{C_*(X,A)}$ is defined by means of the short exact sequence

$$\underline{C_*(A)} \rightarrowtail \underline{C_*(X)} \twoheadrightarrow \underline{C_*(X,A)}.$$

Since this is a sequence of projective complexes (always assuming that $\mathfrak{F}$ contains $\mathfrak{F}(X)$), applying $\mathrm{mor}(\,?\,, M)$ yields a short exact sequence of cochain complexes of abelian groups which, by taking cohomology gives rise in the usual way to a long exact cohomology sequence

$$\cdots \to H^i_{\mathfrak{F}}(X;M) \to H^i_{\mathfrak{F}}(A;M) \to H^{i+1}_{\mathfrak{F}}((X,A);M) \to H^{i+1}_{\mathfrak{F}}(X;M) \to \cdots .$$

If X is a G-CW-complex, one can use the cellular homology groups to define contravariant functors $\underline{H_i(X)}$ by putting

$$\underline{H_i(X)}(G/K) = H_i(X^K;\mathbb{Z}).$$

The reader is invited to check that the category $\mathrm{Mod}_{\mathfrak{F}}$-$G$ has enough injectives, so that every $M \in \mathrm{Mod}_{\mathfrak{F}}$-$G$ admits an injective coresolution $M \rightarrowtail I_*(M)$. By filtering the double complex $\mathrm{mor}(\underline{C_*(X)}, I_*(M))$ suitably, one obtains a *universal coefficient spectral sequence*, which takes the form

$$E_2^{i,j} = \mathrm{Ext}^i(\underline{H_j(X)}, M) \Longrightarrow H^{i+j}_{\mathfrak{F}}(X;M),$$

where again $\mathfrak{F}$ is supposed to contain $\mathfrak{F}(X)$.

A G-homotopy equivalence between G-CW-complexes $f : X \to Y$ induces homotopy equivalences $X^H \to Y^H$ and therefore an isomorphism $\underline{H_*(X)} \cong \underline{H_*(Y)}$. Using the naturality of the above spectral sequence, this shows that $H^*_{\mathfrak{F}}(X; M)$ depends on the G-homotopy type of X only. (Alternatively, one could check that $\underline{C_*(f)} : \underline{C_*(X)} \to \underline{C_*(Y)}$ is a chain homotopy equivalence, by observing that the associated mapping cone is contractible.) It follows that if X is G-homotopy equivalent to a G-CW-complex of dimension n then $H^i_{\mathfrak{F}}(X; M)$ is zero for all $i > n$ and all $M \in \mathrm{Mod}_{\mathfrak{F}}$-$G$. This condition is easily seen to be equivalent to the condition that $\underline{C_*(X)}$ is chain homotopy equivalent to a projective complex of length $\leq n$. We will make use of the following notation.

Definition 3.6. Let X be a G-CW-complex and assume that $\mathfrak{F}(X) \subset \mathfrak{F}$. Then $\mathrm{cd}_{\mathfrak{F}} X \leq n$ if $H^i_{\mathfrak{F}}(X; ?) = 0$ for all $i > n$.

The following theorem is due to Lück [83].

Theorem 3.7. *Let X be a G-CW-complex with* $\mathrm{cd}_{\mathfrak{F}} X \leq n$ *and assume that* $\mathfrak{F}(X) \subset \mathfrak{F}$. *Then X is G-homotopy equivalent to a G-CW-complex of dimension* $\max(3, n)$.

Corollary 3.8. *The universal space $E\mathfrak{F}$ is G-homotopy equivalent to a finite dimensional G-CW-complex if and only if $H^i_{\mathfrak{F}}(G; ?) = 0$ for $i >> 0$.*

For more results concerning the dimension of $\underline{E}G$ see Nucinkis [104], Lück [87], Lück–Meintrup [92], and [101], [73].

One can also use homotopy groups to define contravariant functors $\underline{\pi_i}$. To make sure that the values are abelian groups and that we do not have problems arising from choosing base points, we restrict our attention to *G-simple G-CW-complexes*, defined as follows. Recall that a connected CW-complex X is called *simple* if $\pi_1(X)$ is abelian and operates trivially on the higher homotopy groups of X.

Definition 3.9. A G-CW-complex X is called *G-simple* if X^H is connected and simple for every $H \in \mathfrak{F}(X)$.

For instance, if $\mathfrak{F}$ denotes a family of subgroups of G which is closed under conjugation and taking subgroups, then the universal space $E\mathfrak{F}$ is G-simple. If X is a G-simple G-CW-complex, one obtains objects $\underline{\pi_i(X)}$ in $\mathrm{Mod}_{\mathfrak{F}}$-$G$ by putting $\underline{\pi_i(X)}(G/K)$ equal to $\pi_i(X^K)$ (homotopy groups of the empty space are put equal to $\{e\}$).

Similarly to the non-equivariant case, given a G-CW-pair (X, A) one can define obstructions to existence and uniqueness for extending (up to G-homotopy) a G-map $f_A : A \to Y$ to a G-map $X \to Y$. Given an extension of f_A to a G-map $f : X^n \cup A \to Y$ one obtains, by composing f with the attaching map $S^n \times \underline{\Delta_{n+1}} \to X^n$ of the $(n+1)$-cells, an obstruction cocycle $c_f \in \mathrm{mor}(\underline{C_{n+1}(X, A)}, \underline{\pi_n(Y)})$ which vanishes if and only if f extends to $X^{n+1} \cup A$, yielding the following theorem (cf. [96] for more details).

Theorem 3.10. **(cf.** [96]**)** *Let (X, A) be a G-CW-pair and Y a G-simple G-CW-complex with $\mathfrak{F}(X) \subset \mathfrak{F}(Y) \subset \mathfrak{F}$. Then the restriction of $f : X^n \cup A \to Y$ to $X^{n-1} \cup A$ extends to a map $g : X^{n+1} \cup A \to Y$ if and only if $[c_f] \in H^{n+1}_{\mathfrak{F}}(X, A; \underline{\pi_n(Y)})$ vanishes. Moreover, the restriction of g to $X^n \cup A$ is G-homotopic to f rel. $X^{n-2} \cup A$ if and only if a certain obstruction in $H^n_{\mathfrak{F}}(X, A; \underline{\pi_n(Y)})$ vanishes.*

As an application one can deduce the universal property of $\underline{E\mathfrak{F}}$ (cf. Theorem 2.4) as follows.

Corollary 3.11. *Let $\mathfrak{F}$ be a family of subgroups of G, closed under conjugation and passing to subgroups. Let X be a G-CW-complex with $\mathfrak{F}(X) \subset \mathfrak{F}$. Then there is a unique G-homotopy class of maps*

$$X \longrightarrow E\mathfrak{F}.$$

Proof. Let $K \in \mathfrak{F}(X)$. A G-map $G/K \to E\mathfrak{F}$ corresponds to a point in the contractible space $(E\mathfrak{F})^K$ and, therefore, there is a unique G-homotopy class $G/K \to E\mathfrak{F}$. Since X^0 is a disjoint union of such orbits, there is a unique G-homotopy class $\mu : X^0 \to E\mathfrak{F}$. Since $\underline{\pi_i(E\mathfrak{F})} = 0$ for $i \geq 0$, all obstructions to extending μ uniquely over X vanish. □

To define Bredon *homology* groups, we need to consider *covariant* functors $N : \mathfrak{O}_{\mathfrak{F}}(G) \to \mathfrak{Ab}$, which form an abelian category denoted by G-$\mathrm{Mod}_{\mathfrak{F}}$ (recall that we denote the *contravariant* functors $\mathfrak{O}_{\mathfrak{F}}(G) \to \mathfrak{Ab}$ by $\mathrm{Mod}_{\mathfrak{F}}$-$G$). The definition of

$$\mathrm{Tor}_i(M, N), \quad M \in \mathrm{Mod}_{\mathfrak{F}}\text{-}G \quad \text{and} \quad N \in G\text{-}\mathrm{Mod}_{\mathfrak{F}}$$

is as follows: $\mathrm{Tor}_i(\,?\,, N)$ is the i-th left derived functor of the *categorical tensor product* functor $(\,?\,) \otimes_{\mathfrak{F}} N : \mathrm{Mod}_{\mathfrak{F}}\text{-}G \to \mathfrak{Ab}$. Per definition, $M \otimes_{\mathfrak{F}} N$ is the abelian group $\sum M(G/K) \otimes N(G/K)/\sim$, where the equivalence relation is generated by $\phi^* m \otimes n \sim m \otimes \phi_* n$, with $\phi \in \mathrm{mor}(G/K, G/L)$ and $m \in M(G/L), n \in N(G/K)$. For instance, if $M = P_L$, the projective module defined earlier, then $P_L \otimes_{\mathfrak{F}} N = N(G/L)$. Clearly, the tensor product commutes with coproducts. Since a general projective P in $\mathrm{Mod}_{\mathfrak{F}}$-$G$ is a direct summand of a projective of the form $\bigoplus P_{K_\alpha}$, the functor $P \otimes_{\mathfrak{F}} (\,?\,)$ is exact for any projective P. This implies that the groups $\mathrm{Tor}_i(M, N)$ can be computed as the homology groups of the chain complex $P_*(M) \otimes_{\mathfrak{F}} N$, where $P_*(M)$ is a projective resolution of M. In computations it is useful to know the adjunction relation

$$\mathrm{Hom}(M \otimes_{\mathfrak{F}} N, A) \cong \mathrm{mor}(M, \mathrm{Hom}(N, A))$$

where $M \in \mathrm{Mod}_{\mathfrak{F}}$-$G$, $N \in G$-$\mathrm{Mod}_{\mathfrak{F}}$ and $A \in \mathfrak{Ab}$; $\mathrm{Hom}(N, A)$ stands for the contravariant functor $\mathfrak{O}_{\mathfrak{F}}(G) \to \mathfrak{Ab}$ with

$$\mathrm{Hom}(N, A)(G/K) = \mathrm{Hom}_{\mathbb{Z}}(N(G/K), A).$$

Definition 3.12. The *Bredon homology groups of G with coefficients in* $N \in G$-$\mathrm{Mod}_{\mathfrak{F}}$ are given by

$$H_i^{\mathfrak{F}}(G; N) = \mathrm{Tor}_i(\underline{\mathbb{Z}}, N), \quad i \geq 0.$$

For instance, $H_0^{\mathfrak{F}}(G;N) = \underline{\mathbb{Z}} \otimes_{\mathfrak{F}} N = \operatorname{colim}_{G/K \in \mathcal{O}_{\mathfrak{F}}(G)} N(G/K)$.

Definition 3.13. If X is a G-CW-complex and $N \in G\text{-Mod}_{\mathfrak{F}}$, with $\mathfrak{F}$ a family containing the isotropy groups $\mathfrak{F}(X)$ of the G-action on X, then

$$H_i^{\mathfrak{F}}(X;N) = H_i(\underline{C_*(X)} \otimes_{\mathfrak{F}} N), \quad i \geq 0.$$

As in the case of Bredon cohomology groups, the definition yields groups independent of $\mathfrak{F}$, assuming that $\mathfrak{F}(X) \subset \mathfrak{F}$. The functor $H_*^{\mathfrak{F}}(G;\,?\,)$ is equivalent to $H_*^{\mathfrak{F}}(E\mathfrak{F};\,?\,)$, since $\underline{C_*(E\mathfrak{F})} \to \underline{\mathbb{Z}}$ is a projective resolution. Short exact sequences in $G\text{-Mod}_{\mathfrak{F}}$ give rise to long exact homology sequences, because $P \otimes_{\mathfrak{F}} (\,?\,)$ is an exact functor for P projective. Homology groups for pairs of G-CW-complexes are defined in the obvious way, and one obtains associated long exact homology sequences. The functors $H_*^{\mathfrak{F}}(X;\,?\,)$ depend on the G-homotopy type of X only. The following is a useful example.

Lemma 3.14. *Suppose that $E\mathfrak{F}$ is G-homotopy equivalent to a 1-dimensional G-CW-complex X. Write S_e for the stabilizer of an edge e and S_v for the stabilizer of a vertex v of X. Let $N \in G\text{-Mod}_{\mathfrak{F}}$. Then $H_i^{\mathfrak{F}}(E\mathfrak{F};N) = 0$ for $i > 1$ and there is an exact sequence*

$$H_1^{\mathfrak{F}}(E\mathfrak{F};N) \rightarrowtail \bigoplus_{[e]} N(G/S_e) \to \bigoplus_{[v]} N(G/S_v) \twoheadrightarrow H_0^{\mathfrak{F}}(E\mathfrak{F};N)$$

where $[e]$ resp. $[v]$ runs over the G-orbits of edges resp. vertices of X.

Proof. The chain complex of X gives rise to an exact sequence in $\text{Mod}_{\mathfrak{F}}\text{-}G$ of the form

$$\bigoplus_{[e]} P_{S_e} \rightarrowtail \bigoplus_{[v]} P_{S_v} \twoheadrightarrow \underline{\mathbb{Z}}.$$

Applying $\operatorname{Tor}_*(\,?\,,N)$ to this short exact sequence yields the long exact sequence of the lemma. □

For instance, if $G = H *_K L$ is an amalgamated product and $\mathfrak{F} = \mathfrak{F}[H,L]$ denotes the smallest family of subgroups of G, which contains H and L and which is closed under conjugation and passing to subgroups, then $E\mathfrak{F}[H,L]$ is G-homotopy equivalent to a 1-dimensional G-CW-complex X. Indeed, G acts on a tree X with edge stabilizers conjugate to K and vertex stabilizers conjugate to H and L respectively so that the orbit space X/G is an interval, with the two vertices corresponding to H and L and the edge corresponding to K (cf. Serre's book [122]).

Corollary 3.15. *Let $G = H *_K L$ and $\mathfrak{F} = \mathfrak{F}[H,L]$. Then $H_i^{\mathfrak{F}[H,L]}(G;N) = 0$ for $i > 1$ and there is a natural exact sequence, where the middle arrow is induced by the projections of G/K onto G/H and G/L respectively:*

$$H_1^{\mathfrak{F}[H,L]}(G;N) \rightarrowtail N(G/K) \to N(G/H) \oplus N(G/L) \twoheadrightarrow H_0^{\mathfrak{F}[H,L]}(G;N).$$

Remark 3.16. If one is interested in Bredon (co)homology for arbitrary (not necessarily CW) G-spaces X, one can define *singular* Bredon (co)homology in the following obvious way. One applies the previous construction to $\Gamma(X)$, the geometric realization functor mentioned earlier (cf. Remark 2.3). This always yields a simplicial space with simplicial G-action G-homeomorphic to a G-CW-complex (as one sees by passing to the second barycentric subdivision). Note that if the G-action on X has finite stabilizers, then $\Gamma(X)$ is G-homeomorphic to a proper G-CW-complex. Clearly, any weak G-homotopy equivalence $f: X \to Y$ induces a G-homotopy equivalence $\Gamma(X) \to \Gamma(Y)$ and therefore X and Y will have the same singular Bredon (co)homology groups. Also, if X is already a G-CW-complex, then the canonical map $\Gamma(X) \to X$ is a G-homotopy equivalence by the equivariant Whitehead Theorem 2.2, showing that in this case the Bredon (co)homology of X agrees with its singular Bredon (co)homology.

The following example plays a role in connection with the Baum–Connes Conjecture 5.9. We write $\mathfrak{Fin}(G)$ for the class of finite subgroups of a group G. If the group G is clear from the context, we just write $\mathfrak{Fin}$. As earlier, we will use the notation $\underline{E}G$ for the universal space $E\mathfrak{Fin}(G)$. We will also make use of the notation

$$H^i_{\mathfrak{Fin}}(G; ?) \quad \text{and} \quad H^i_{\mathfrak{Fin}}(X; ?)$$

for the Bredon cohomology with respect to the family of finite subgroups of the group G respectively the proper G-CW-complex X; similarly for Bredon homology. For any group G we define $R^\sharp \in \mathrm{Mod}_{\mathfrak{Fin}}\text{-}G$ and $R_\sharp \in G\text{-}\mathrm{Mod}_{\mathfrak{Fin}}$ by

$$R^\sharp(G/H) = R_\sharp(G/H) = \mathbb{Q} \otimes R_{\mathbb{C}}(H),$$

where $R_{\mathbb{C}}(H)$ denotes the underlying abelian group of the complex representation ring of the finite group H. For the morphisms, the definition is as follows. A morphism $f: G/H \to G/K$ with $f(H) = gK$ gives rise to a group homomorphism $H \to K$, $h \mapsto g^{-1}hg$, which is unique up to conjugation in K: if $f(H) = g_1K = g_2K$ then $g_1 = g_2k$ for some $k \in K$ so that $g_1^{-1}hg_1 = k^{-1}(g_2^{-1}hg_2)k$. Using restriction (resp. induction) on the level of representations and because inner automorphisms act trivially on representation rings, one thus has induced maps of abelian groups

$$f^\sharp : R^\sharp(K) \to R^\sharp(H); \quad f_\sharp : R_\sharp(H) \to R_\sharp(K).$$

The contravariant and covariant nature of the functor $\mathbb{Q} \otimes R_{\mathbb{C}}(?)$ are linked via its *Mackey functor structure* (see [88]), which we won't discuss here. In the sequel, we just write $\mathbb{Q} \otimes R_{\mathbb{C}}$ for $R_\sharp$ respectively $R^\sharp$, if it is clear from the context whether $\mathbb{Q} \otimes R_{\mathbb{C}}$ is to be considered as a covariant or a contravariant functor. Similarly, we write $R_{\mathbb{C}}$ for the (contravariant, respectively covariant) functor $G/H \mapsto R_{\mathbb{C}}(H)$.

In the context of the *Baum–Connes Conjecture*, we will particularly be interested in the groups

$$H_i^{\mathfrak{Fin}}(\underline{E}G; R_{\mathbb{C}}), \quad i \geq 0.$$

In case G admits a one-dimensional $\underline{E}G$, we can apply Lemma 3.14 and we obtain an explicit result as follows.

Theorem 3.17. *Suppose G admits a 1-dimensional model for $\underline{E}G$, say T. Then*

$$H_i^{\mathfrak{Fin}}(G; R_{\mathbb{C}}) = 0, \quad \text{if} \quad i > 1,$$

and there is a short exact sequence

$$H_1^{\mathfrak{Fin}}(G; R_{\mathbb{C}}) \rightarrowtail \bigoplus_{[e]} R_{\mathbb{C}}(S_e) \to \bigoplus_{[v]} R_{\mathbb{C}}(S_v) \twoheadrightarrow H_0^{\mathfrak{Fin}}(G; R_{\mathbb{C}}).$$

The groups S_e and S_v denote the stabilizers of the edge e respectively vertex v of T and the sums are taken over the G-orbits of edges and vertices respectively.

We recall the well-known fact that the class of groups G admitting a one-dimensional $\underline{E}G$ coincides with the class of groups satisfying $\mathrm{cd}_{\mathbb{Q}}\, G \leq 1$ (cf. Dunwoody [40]).

It is convenient to extend the definition of the representation ring to arbitrary groups by taking a colimit over finite subgroups as follows.

Definition 3.18. Let G be an arbitrary group. Then

$$R_{\mathbb{C}}^{\mathfrak{Fin}}(G) := H_0^{\mathfrak{Fin}}(\underline{E}G; R_{\mathbb{C}}) = \mathrm{colim}_{G/H \in \mathfrak{O}_{\mathfrak{Fin}}(G)} R_{\mathbb{C}}(H).$$

Note that in case G has only finitely many conjugacy classes of finite subgroups, $R_{\mathbb{C}}^{\mathfrak{Fin}}(G)$ is a finitely generated abelian group, being a factor group of a finite sum of the form $\bigoplus R_{\mathbb{C}}(H)$, where H runs over the conjugacy classes of finite subgroups of G; as we will see (cf. Theorem 3.25), the rank of the abelian group $R_{\mathbb{C}}^{\mathfrak{Fin}}(G)$ is in this case equal to the number of conjugacy classes of elements of finite order in G. Indeed, for an arbitrary group G the following holds. We write $\mathbb{C}[\mathrm{FC}(G)]$ for the $\mathbb{C}$-vector space spanned by the set of conjugacy classes $\mathrm{FC}(G)$ of elements of finite order in G. Recall that for a finite group H, the irreducible $\mathbb{C}$-representations form a basis of the free abelian group $R_{\mathbb{C}}(H)$ and their characters form a basis of the $\mathbb{C}$-vector space of class functions $H \to \mathbb{C}$. One thus has a natural isomorphism

$$\chi_H : R_{\mathbb{C}}(H) \otimes \mathbb{C} \xrightarrow{\cong} \mathbb{C}[\mathrm{FC}(H)]$$

associating to a complex linear combination of representations the values of the associated sum of characters on each conjugacy class. A similar result holds for an arbitrary group as follows.

Theorem 3.19. *Let G be an arbitrary group and write $\mathrm{FC}(G)$ for the set of conjugacy classes of elements of finite order in G. Then there is a natural map induced by the Hattori–Stallings trace*

$$HS_G^{\mathfrak{Fin}} : R_{\mathbb{C}}^{\mathfrak{Fin}}(G) \longrightarrow \mathbb{C}[\mathrm{FC}(G)]$$

inducing, upon tensoring the domain by $\mathbb{C}$, *an isomorphism*

$$HS^{\mathfrak{Fin}}_{G,\mathbb{C}} : R^{\mathfrak{Fin}}_{\mathbb{C}}(G) \otimes_{\mathbb{Z}} \mathbb{C} \xrightarrow{\cong} \mathbb{C}[\mathrm{FC}(G)].$$

Proof. If H is a finite group and P is a finitely generated $\mathbb{C}[H]$-module, then the Hattori–Stallings trace (cf. Hattori [57] and Stallings [126])

$$HS_H(P) = \sum_{[h]\in FC(H)} HS_H(P)(h)\cdot[h] \in \mathbb{C}[\mathrm{FC}(H)], \quad HS_H(P)(h) \in \mathbb{C},$$

is defined as the trace of a projection $\mathbb{C}[H]^n \to \mathbb{C}[H]^n$ with image isomorphic to P, with this trace to be considered as an element in $\mathbb{C}[\mathrm{FC}(H)]$ to make it independent of choices (cf. Appendix Lemma 7.13 for more details). The complex coefficient $HS_H(P)(h)$ of the conjugacy class $[h]$ of $h \in H$ can be expressed in terms of the character of P by the formula

$$HS_H(P)(h) = \frac{1}{|C_H(h)|}\cdot \chi_H(P)(h^{-1}),$$

where $C_H(h)$ denotes the centralizer of h (see also Bass [3]). This implies that for H finite, the induced homomorphism

$$HS_{H,\mathbb{C}} : R_{\mathbb{C}}(H)\otimes\mathbb{C} \longrightarrow \mathbb{C}[\mathrm{FC}(H)]$$

is an isomorphism of $\mathbb{C}$-vector spaces, since HS_H agrees with the character map χ_H up to an automorphism of the group $R_{\mathbb{C}}(H)\otimes\mathbb{C}$. Clearly, if $\alpha : H \to K$ is any homomorphism of finite groups, one has a commutative diagram

$$\begin{array}{ccc} R_{\mathbb{C}}(H) & \xrightarrow{\mathrm{Ind}_\alpha} & R_{\mathbb{C}}(K) \\ {\scriptstyle HS_H}\downarrow & & \downarrow{\scriptstyle HS_K} \\ \mathbb{C}[\mathrm{FC}(H)] & \xrightarrow{\alpha_*} & \mathbb{C}[\mathrm{FC}(K)] \end{array}$$

because the elements a_{ij} of a matrix describing the projection $\mathbb{C}[H]^n \to \mathbb{C}[H]^n$ with image P turn into the elements $\alpha(a_{ij})$ under the map $P \mapsto \mathrm{Ind}_\alpha(P)$. This compatibility with induction is the reason that, in the course of this proof, we use the Hattori–Stallings trace rather than the character map χ_H discussed before; χ_H behaves well under restriction and would be appropriate in the study of the *contravariant* version of $R_{\mathbb{C}}$. By taking the direct limit we define the map

$$HS^{\mathfrak{Fin}}_{G} := \mathrm{colim}_{G/H\in\mathfrak{O}_{\mathfrak{Fin}}}(HS_H : R_{\mathbb{C}}(H) \to \mathbb{C}[\mathrm{FC}(H)])$$

and, as each map $HS_{H,\mathbb{C}} : R_{\mathbb{C}}(H)\otimes\mathbb{C} \to \mathbb{C}[\mathrm{FC}(H)]$ is an isomorphism, $HS^{\mathfrak{Fin}}_{G,\mathbb{C}} := \mathrm{colim}\, HS_{H,\mathbb{C}}$ is an isomorphism too and the claim of the theorem follows by observing that

$$\mathbb{C}[\mathrm{FC}(G)] \cong \mathrm{colim}_{G/H\in\mathfrak{O}_{\mathfrak{Fin}}}\, \mathbb{C}[\mathrm{FC}(H)].$$

□

Remark 3.20. Using only the component of the neutral element $e \in G$ in the definition of the Hattori–Stallings trace, and because for a finite dimensional representation P of a finite group H one has $HS_H(P)(e) = \dim_{\mathbb{C}}(P)/|H|$, the map $HS_G(?)(e) =: \kappa$ takes its values in $\mathbb{Q}$. We call the map

$$\kappa : R_{\mathbb{C}}^{\mathfrak{Fin}}(G) \to \mathbb{Q}$$

the *Kaplansky* trace; it extends to the usual Kaplansky trace on the projective class group (cf. Appendix), fitting into a commutative diagram

$$\begin{array}{ccc} R_{\mathbb{C}}^{\mathfrak{Fin}}(G) & \longrightarrow & K_0^{alg}(\mathbb{C}[G]) \\ {\scriptstyle\kappa}\downarrow & & \downarrow{\scriptstyle\kappa_{\mathbb{C}}} \\ \mathbb{Q} & \xrightarrow{=} & \mathbb{Q}. \end{array}$$

It is obvious that $\kappa(R_{\mathbb{C}}^{\mathfrak{Fin}}(G)) \subset \mathbb{Q}$ is the subgroup generated by the elements $1/|H|$, where $H < G$ runs through the finite subgroups; for $\mathrm{Im}(\kappa_{\mathbb{C}})$ this is a conjecture, see **PCGC** on page 70 (more on the Kaplansky trace, in particular a proof of Zalesskii's Theorem that $\kappa_{\mathbb{C}}$ takes values in $\mathbb{Q}$, can be found in Burger–Valette [19]).

For the higher groups $H_i^{\mathfrak{Fin}}(\underline{E}G; R_{\mathbb{C}})$ the following obvious result is useful to record. We denote by $(\underline{E}G)^{\mathrm{sing}}$ the *singular* part of $\underline{E}G$, that is, the subcomplex consisting of all points in $\underline{E}G$ with non-trivial isotropy group. Clearly, $(\underline{E}G)^{\mathrm{sing}}$ is up to G-homotopy uniquely determined by G. We write $\dim(\underline{E}G)^{\mathrm{sing}}$ for the minimal dimension of $(\underline{E}G)^{\mathrm{sing}}$ within its G-homotopy type. We also note that if $(\underline{E}G)^{\mathrm{sing}}$ is G-homotopy equivalent to the G-CW-complex A then there exists a model X for $\underline{E}G$ such that $X^{\mathrm{sing}} = A$. Namely, we can construct X as $\cup X(n)$ where $X(n) = A * G(n)$ with $G(n)$ an n-fold join of copies of G. Since the functor $R_{\mathbb{C}}$ takes the value $\mathbb{Z}$ on orbits $G/\{e\}$, with no G-action on $\mathbb{Z}$, the map of complexes induced by the projection $\underline{E}G \to (\underline{E}G)/G$

$$\underline{C_*(\underline{E}G)} \otimes_{\mathfrak{F}} R_{\mathbb{C}} \to C_*(\underline{E}G) \otimes_G \mathbb{Z}$$

is an isomorphism in degree $> \dim \underline{E}G^{\mathrm{sing}}$ and surjective in degree $\dim \underline{E}G^{\mathrm{sing}}$, so that the following holds. We write $\underline{B}G$ for the orbit space $(\underline{E}G)/G$ which, for a torsion-free group G, is homotopy equivalent to $K(G, 1)$.

Lemma 3.21. *Let G be an arbitrary group. Then there is a natural map*

$$H_i^{\mathfrak{Fin}}(\underline{E}G; R_{\mathbb{C}}) \longrightarrow H_i(\underline{B}G; \mathbb{Z})$$

which is an isomorphism in dimensions $i > \dim(\underline{E}G)^{\mathrm{sing}} + 1$ and injective in dimension $i = \dim(\underline{E}G)^{\mathrm{sing}} + 1$.

In case the singular part is zero-dimensional, one can discuss $H_0^{\mathfrak{Fin}}$ and $H_1^{\mathfrak{Fin}}$ more precisely, as the following lemma shows; we will make use of this later when we discuss the Baum–Connes Conjecture for one-relator groups.

Lemma 3.22. *If $\underline{E}G^{sing}$ is G-homotopy equivalent to a single orbit G/H then G has up to conjugation a unique maximal finite subgroup and this subgroup is conjugate to H. Moreover*

$$H_0^{\mathfrak{Fin}}(G; R_{\mathbb{C}}) \cong R_{\mathbb{C}}(H)$$

and

$$H_1^{\mathfrak{Fin}}(G; R_{\mathbb{C}}) \cong H_1(\underline{B}G; \mathbb{Z}) \cong (G/Tor(G))_{ab},$$

where $Tor(G)$ denotes the subgroup of G generated by the torsion elements of G.

Proof. By choosing an appropriate model, we can assume that $\underline{E}G^{\text{sing}} = G/H$. Let $F < G$ be a finite subgroup $\neq \{e\}$. Then the assumption on the singular part of $\underline{E}G$ implies that $(G/H)^F$ is discrete and contractible, thus it is a point. Therefore, F is conjugate to a subgroup of H. It follows that H is a maximal finite subgroup of G, and every maximal finite subgroup of G is conjugate to H. Since in particular $(G/H)^H \cong N_G(H)/H = \{*\}$, we see that $N_G(H) = H$ and therefore

$$R_{\mathbb{C}}^{\mathfrak{Fin}}(G) = H_0^{\mathfrak{Fin}}(G; R_{\mathbb{C}}) = \operatorname{colim}_{G/F \in \mathfrak{O}_{\mathfrak{Fin}}(G)} R_{\mathbb{C}}(F) \cong R_{\mathbb{C}}(H),$$

proving the first part. It is a well-known fact that in case X is a 1-connected G-CW-complex, $\pi_1(X/G) = G/Stab(G)$ where $Stab(G)$ denotes the normal subgroup of G normally generated by the stabilizers of the G-action on X (cf. Armstrong [1]). In particular $\pi_1(\underline{B}G) = G/Tor(G)$, and to finish our proof we only need to show that the map $H_1^{\mathfrak{Fin}}(\underline{E}G; R_{\mathbb{C}}) \to H_1(\underline{B}G; \mathbb{Z})$, which is injective by Lemma 3.21, is surjective too. If we write $\underline{C_i}$ for $\underline{C_i(\underline{E}G) \otimes_{\mathfrak{F}} R_{\mathbb{C}}}$ and C_* for the ordinary cellular chain complex $C_*(\underline{B}G)$, then

$$\underline{C_0} \cong (\bigoplus_J \mathbb{Z}) \oplus R_{\mathbb{C}}(H), \quad C_0 = (\bigoplus_J \mathbb{Z}) \oplus \mathbb{Z}.$$

Here J denotes the set of free orbits of vertices. A simple diagram chase in the diagram

$$\begin{array}{ccccccc}
 & & & & \operatorname{Ker}(\pi) & \xrightarrow{\cong} & \tilde{R}_{\mathbb{C}}(H) \\
 & & & & \downarrow{\scriptstyle \text{mono}} & & \downarrow{\scriptstyle \text{mono}} \\
\underline{C_2} & \xrightarrow{\underline{d_2}} & \underline{C_1} & \xrightarrow{\underline{d_1}} & \underline{C_0} & \xrightarrow{\text{epi}} & R_{\mathbb{C}}(H) \\
\downarrow{\scriptstyle =} & & \downarrow{\scriptstyle =} & & \downarrow{\scriptstyle \pi} & & \downarrow{\scriptstyle \epsilon} \\
C_2 & \xrightarrow{d_2} & C_1 & \xrightarrow{d_1} & C_0 & \xrightarrow{\text{epi}} & \mathbb{Z}
\end{array}$$

with $\pi = (\mathrm{Id}, \epsilon)$, and $\epsilon : R_C(H) \to \mathbb{Z}$ the augmentation. The diagram shows that $\operatorname{Ker} \underline{d_1} = \operatorname{Ker} d_1$, thus $\operatorname{Ker} \underline{d_1} / \operatorname{Im} \underline{d_2} \twoheadrightarrow \operatorname{Ker} d_1 / \operatorname{Im} d_2$. □

If G is a one-relator group, there exists a two-dimensional model for $\underline{E}G$ such that, in case G is not torsion-free, the singular part consists of a single orbit G/C, $C < G$ a finite cyclic subgroup (see Brown's book [18, Chapter II, Section 5]). As we have seen in the course of the proof of the previous Lemma, this implies that up to conjugation, G possesses a unique maximal finite subgroup and this subgroup is conjugate to C; also, its normalizer $N_G(C)$ satisfies $N_G(C) = C$. (The reader should consult Lyndon–Schupp [94] for facts concerning one relator groups.)

Corollary 3.23. *Let $G = \langle x_\alpha;\ \alpha \in I \,|\, r\rangle$ be a one relator group with maximal finite subgroup $C < G$. Then $H_i^{\mathfrak{Fin}}(G; R_{\mathbb{C}}) = 0$ for $i > 2$ and*

$$H_0^{\mathfrak{Fin}}(G; R_{\mathbb{C}}) \cong R_{\mathbb{C}}(C), \quad H_1^{\mathfrak{Fin}}(G; R_{\mathbb{C}}) \cong G_{ab}/\overline{C}$$

where $\overline{C}$ denotes the image of C in G_{ab}. Moreover,

$$H_2^{\mathfrak{Fin}}(G; R_{\mathbb{C}}) \cong H_2(G;\mathbb{Z}) = \begin{cases} \mathbb{Z}, & \text{if } r \in [F,F] \\ 0, & \text{if } r \notin [F,F] \end{cases}$$

where $[F,F]$ denotes the commutator subgroup of the free group F generated by the x_α's.

Proof. The presentation complex X is a two-dimensional simplicial CW-complex whose barycentric subdivision is a model for $\underline{E}G$ (see Brown's book loc. cit. for a description of X). The singular part of X is discrete, of the form G/C so we can read off the part concerning $H_0^{\mathfrak{Fin}}$, $H_1^{\mathfrak{Fin}}$ and $H_i^{\mathfrak{Fin}}$, $i > 2$, from the previous Lemma and Lemma 3.21. The cellular chain complex $C_*(X)$ has the form

$$C_2(X) = \mathbb{Z}[G/C] \to C_1(X) = \bigoplus_I \mathbb{Z}[G] \to C_0(X) = \mathbb{Z}[G] \to \mathbb{Z}$$

which, because X is acyclic, shows that

$$H_2(G;\mathbb{Z}) \cong H_2((\underline{E}G)/G;\mathbb{Z})$$

since both these groups agree with the kernel of the induced map

$$(\mathbb{Z}[G/C] \to \bigoplus_I \mathbb{Z}[G]) \otimes_{\mathbb{Z}[G]} \mathbb{Z}.$$

Applying Lemma 3.21 shows that

$$H_2^{\mathfrak{Fin}}(G; R_{\mathbb{C}}) \cong H_2(G;\mathbb{Z})$$

and from Hopf's celebrated formula [18]

$$H_2(G;\mathbb{Z}) \cong R \cap [F,F]/[F,R],$$

we see that for R the subgroup of F normally generated by the relator r, $H_2(G;\mathbb{Z})$ is as claimed. □

We now return to the problem of determining $H_*^{\mathfrak{Fin}}(G; \mathbb{Q} \otimes R_{\mathbb{C}})$. For the convenience of the reader we recall the following well-known fact on representations of finite cyclic groups.

Lemma 3.24. *Let C_n be a cyclic group of order n. The rationalized complex representation ring of C_n is isomorphic to a product of cyclotomic fields*

$$\mathbb{Q} \otimes R_{\mathbb{C}}(C_n) \cong \prod_{d|n} \mathbb{Q}(\zeta_d),$$

where ζ_d denotes a primitive d-th root of 1, and d runs over the divisors of n. Moreover, the idempotent e_{C_n} of $\mathbb{Q} \otimes R_{\mathbb{C}}(C_n)$ corresponding to $\mathbb{Q}(\zeta_n)$ has the following property. If $C_n < G$ and $Q = N_G(C_n)/C_G(C_n)$, then the fixed-point space

$$(e_{C_n}(\mathbb{Q} \otimes R_{\mathbb{C}}(C_n)))^Q = \bigoplus \mathbb{Q}$$

is a $\mathbb{Q}$-vector space of dimension equal to the number of G-conjugacy classes of elements of order n in C_n, a number which equals $\varphi(n)/|Q|$.

Proof. One first observes that as a ring,

$$\mathbb{Q} \otimes R_{\mathbb{C}}(C_n) \cong \mathbb{Q}[C_n] \cong \mathbb{Q}[x]/(x^n - 1).$$

Factoring $x^n - 1$ into $\mathbb{Q}$-irreducible polynomials yields the decomposition of the representation ring into cyclotomic fields. By identifying the Galois group $G(n)$ of $\mathbb{Q}(\zeta_n)$ over $\mathbb{Q}$ with the automorphism group $(\mathbb{Z}/n\mathbb{Z})^\times$ of C_n, we can view Q as a subgroup of $G(n)$. Since $\mathbb{Q}(\zeta_n)$ is a free $\mathbb{Q}[G(n)]$-module of rank one, it is free as $\mathbb{Q}[Q]$-module, of rank equal the index $[G(n) : Q] = \varphi(n)/|Q|$. It follows that the Q-invariant part of $\mathbb{Q}(\zeta_n)$ has dimension $\varphi(n)/|Q|$ as a vector space over $\mathbb{Q}$; by the choice of Q this equals the number of G-conjugacy classes of generators of C_n. □

Theorem 3.25. *Let G be an arbitrary group. Then one has*

$$H^i_{\mathfrak{Fin}}(G; \mathbb{Q} \otimes R_{\mathbb{C}}) = H^i_{\mathfrak{Fin}}(\underline{E}G; \mathbb{Q} \otimes R_{\mathbb{C}}) \cong \prod_{[x] \in \mathrm{FC}(G)} H^i(C_G(x); \mathbb{Q})$$

and

$$H_i^{\mathfrak{Fin}}(G; \mathbb{Q} \otimes R_{\mathbb{C}}) = H_i^{\mathfrak{Fin}}(\underline{E}G; \mathbb{Q} \otimes R_{\mathbb{C}}) \cong \bigoplus_{[x] \in \mathrm{FC}(G)} H_i(C_G(x); \mathbb{Q}).$$

The product (resp. sum) is taken over FC(G)*, the set of conjugacy classes of elements of finite order in G. The right hand sides denote ordinary group (co)homology of the centralizers $C_G(x)$, with constant coefficients $\mathbb{Q}$.*

Proof. We follow the ideas of Lück and Oliver [93, Section 5], which have their root in Slominska's paper [124]. We will concentrate on the case of $\mathbb{Q} \otimes R_{\mathbb{C}} = R^\sharp$; the other case is similar. Let $Z(G)$ denote the set of conjugacy classes of finite, cyclic subgroups of G. If S is a finite cyclic subgroup of G, we write $S \in [S] \in Z(G)$. For a finite subgroup $H < G$ we define the idempotent $e_{S,H} \in \mathbb{Q} \otimes R_{\mathbb{C}}(H)$ to be the restriction of the class function $e_S : G \to \mathbb{C}$, whose value on $g \in G$ is 1 if the subgroup $\langle g \rangle$ generated by g is conjugate to S, and zero otherwise. That $e_S|H$ is indeed a rational linear combination of characters (even of characters of

$\mathbb{Q}$-representations) follows from standard facts on representations of finite groups (cf. [121]). There is a natural splitting

$$R^\sharp = \prod_{[S]\in Z(G)} R^\sharp_S,$$

where $R^\sharp_S$ denotes the contravariant functor given on objects by $G/H \mapsto e_{S,H}(\mathbb{Q}\otimes R_\mathbb{C}(H))$. We therefore obtain a splitting

$$H^i_{\mathfrak{Fin}}(G;\mathbb{Q}\otimes R_\mathbb{C}) \cong \prod_{[S]\in Z(G)} H^i_{\mathfrak{Fin}}(G;R^\sharp_S).$$

Now $R^\sharp_S(G/H)$ is 0 if $[S]$ contains no representative $gSg^{-1} < H$, and in the other case is isomorphic to $\mathbb{Q}(\zeta_{|S|})^N$, with N the normalizer of some $gSg^{-1} < H$, acting via an identification of a generator of gSg^{-1} with $\zeta_{|S|}$. It follows that for any $M \in \mathrm{Mod}_{\mathfrak{Fin}}\text{-}G$

$$\mathrm{mor}(M, R^\sharp_S) \cong \mathrm{Hom}_{N_G(S)}(M(G/S), \mathbb{Q}(\zeta_{|S|}))$$

where $N_G(S)$ acts on $\mathbb{Q}(\zeta_{|S|})$ via an identification of a generator of S with $\zeta_{|S|}$. Therefore,

$$\mathrm{mor}(\underline{C_*(\underline{E}G)}, R^\sharp) \cong \prod_{[S]\in Z(G)} \mathrm{Hom}_{N_G(S)}(C_*(\underline{E}G^S), \mathbb{Q}(\zeta_{|S|}))\,.$$

Recall that for any group K and any finite subgroup $L < K$, the $\mathbb{Q}[K]$-module $\mathbb{Q}[K/L]$ is projective, since it is isomorphic to the induced module $\mathbb{Q}[K]\otimes_L \mathbb{Q}$ and $\mathbb{Q}$ is a projective $\mathbb{Q}[L]$-module. Since, in the notation above, $N_G(S)$ acts properly on the space $\underline{E}G^S$, the complex $C_*(\underline{E}G^S)$ is in each degree $* > 0$ a sum of permutation modules of the form $\mathbb{Z}[N_G(S)/H]$ with H finite. Therefore, using that $\underline{E}G^S$ is contractible, $C_*(\underline{E}G^S)\otimes\mathbb{Q}$ is a $\mathbb{Q}[N_G(S)]$-projective resolution of $\mathbb{Q}$ as a trivial $\mathbb{Q}[N_G(S)]$-module. It follows that

$$H^*(\mathrm{Hom}_{N_G(S)}(C_*(\underline{E}G^S), \mathbb{Q}(\zeta_{|S|}))) \cong H^*(N_G(S);\mathbb{Q}(\zeta_{|S|})).$$

The short exact sequence

$$C_G(S) \rightarrowtail N_G(S) \twoheadrightarrow Q(S),$$

with $Q(S)$ a finite group of order dividing $\varphi(|S|)$, yields a collapsing Serre spectral sequence, with edge isomorphism

$$H^*(N_G(S);\mathbb{Q}(\zeta_{|S|})) \cong H^*(C_G(S);\mathbb{Q}(\zeta_{|S|}))^{Q(S)}.$$

Using the previous lemma, and the fact that taking $Q(S)$-invariants commutes with taking rational homology, we see that

$$H^*(C_G(S);\mathbb{Q}(\zeta_{|S|}))^{Q(S)} \cong \prod_{\varphi(|S|)/|Q(S)|} H^*(C_G(S);\mathbb{Q}).$$

Since every conjugacy class $[S]$ of subgroups of order n in G corresponds to $\varphi(n)/|Q(S)|$ conjugacy classes of elements of order n, the result follows. □

Remark 3.26. The splitting $R_\sharp = \bigoplus R_{\sharp,S}$ (and similarly for $R^\sharp$) can also be used to decompose $H_*^{\mathfrak{F}\mathrm{in}}(X; R_\mathbb{C} \otimes \mathbb{Q})$ (resp. $H^*_{\mathfrak{F}\mathrm{in}}(X; R_\mathbb{C} \otimes \mathbb{Q})$) for an arbitrary proper G-CW-complex X. One finds

$$\underline{C_*(X)} \otimes_{\mathfrak{F}} R_\sharp \cong \bigoplus_{[S]\in Z(G)} C_*(X^S) \otimes_{N_G(S)} R_{\sharp,S}(S)$$

and with $W_G(S) = N_G(S)/C_G(S)$, as $C_G(S)$ acts properly on X^S, the obvious map

$$\bigoplus_{[S]\in Z(G)} C_*(X^S) \otimes_{N_G(S)} R_{\sharp,S}(S) \to \bigoplus_{[S]\in Z(G)} C_*(X^S/C_G(S)) \otimes_{W_G(S)} R_{\sharp,S}(S)$$

is a homology isomorphism, showing that

$$H_i^{\mathfrak{F}\mathrm{in}}(X; R_\mathbb{C} \otimes \mathbb{Q}) \cong \bigoplus_{[S]\in Z(G)} H_i(X^S/C_G(S); \mathbb{Q}) \otimes_{W_G(S)} R_{\sharp,S}(S).$$

We used here that $R_{\sharp,S}(S)$ is a projective $\mathbb{Q}[W_G(S)]$-module, because $W_G(S)$ is a finite group. As we have seen, the $\mathbb{Q}$-vector space

$$R_{\sharp,S}(S)^{W_G(S)} \cong \mathbb{Q} \otimes_{W_G(S)} R_{\sharp,S}(S)$$

has dimension $\phi(|S|)/|Q(S)|$, which is the number of G-conjugacy classes of generators of S. This implies the following.

Theorem 3.27. *Let X be a proper G-CW-complex. In the notation of Remark 3.26 we have*

$$H_i(X^S/C_G(S); \mathbb{Q}) \otimes_{W_G(S)} R_{\sharp,S} \cong \bigoplus_{[g]\in[S;G]} H_i(X^g/C_G(g); \mathbb{Q}),$$

where the sum is taken over the set $[S;G]$ of G-conjugacy classes of generators of the cyclic group S, and there is an isomorphism

$$H_i^{\mathfrak{F}\mathrm{in}}(X; R_\mathbb{C} \otimes \mathbb{Q}) \cong \bigoplus_{[g]\in \mathrm{FC}(G)} H_i(X^g/C_G(g); \mathbb{Q})$$

where the sum is taken over all conjugacy classes $\mathrm{FC}(G)$ *of elements of finite order in G.*

The following is a simple example.

Lemma 3.28. *For $G = S\ell(2,\mathbb{Z})$ one has*

$$H_*^{\mathfrak{F}\mathrm{in}}(S\ell(2,\mathbb{Z}); \mathbb{Q} \otimes R_\mathbb{C}) = H_*^{\mathfrak{F}\mathrm{in}}(\underline{E}S\ell(2,\mathbb{Z}); \mathbb{Q} \otimes R_\mathbb{C}) \cong \begin{cases} 0, & \text{for } * > 0 \\ \mathbb{Q}^8, & \text{for } * = 0. \end{cases}$$

Proof. Since $S\ell(2,\mathbb{Z})$ admits a decomposition of the form $C_4 *_{C_2} C_6$, all finite subgroups of $S\ell(2,\mathbb{Z})$ are conjugate to a subgroup of one of the subfactors C_4 resp. C_6. It follows that the centralizers of finite subgroups have the following form: for $\{e\}$ and C_2 the centralizer is all of $S\ell(2,\mathbb{Z})$, whereas the centralizers of the other finite subgroups are all finite. Since the Mayer–Vietoris sequence of the decomposition of $S\ell(2,\mathbb{Z})$ shows that $H_*(S\ell(2,\mathbb{Z});\mathbb{Q}) = \mathbb{Q}$ and since for a finite group F one has $H_*(F;\mathbb{Q}) = \mathbb{Q}$, each of the centralizers contributes one copy of $\mathbb{Q}$ in degree zero. The conjugacy classes of elements of finite order are: $\{e\}$, the conjugacy class of the (central) element of order 2, two conjugacy classes of elements of order 3, two conjugacy classes of elements of order 4, and two conjugacy classes of elements of order 6, which shows that $H_0^{\mathfrak{Fin}}(\underline{E}S\ell(2,\mathbb{Z}),\mathbb{Q}\otimes R_{\mathbb{C}})$ is an 8-dimensional vector space over $\mathbb{Q}$. □

Remark 3.29. The reader should also compute $H_*^{\mathfrak{Fin}}(S\ell(2,\mathbb{Z});R_{\mathbb{C}})$ using Theorem 3.17 and compare it with the result of the previous Lemma.

In a similar way one proves more generally the following.

Corollary 3.30. *Let $G = H *_L K$ where H and K are finite abelian groups. Then*

$$H_*^{\mathfrak{Fin}}(G;\mathbb{Q}\otimes R_{\mathbb{C}}) = H_*^{\mathfrak{Fin}}(\underline{E}G;\mathbb{Q}\otimes R_{\mathbb{C}}) \cong \begin{cases} 0, & \text{for } * > 0 \\ \mathbb{Q}^{|H|+|K|-|L|}, & \text{for } * = 0. \end{cases}$$

We also could have used Corollary 3.15 to conclude that for (not necessarily abelian) finite groups $H > L < K$ one has for $G = H *_L K$

$$H_i^{\mathfrak{Fin}}(G;R_{\mathbb{C}}) = 0 \quad \text{for} \quad i > 1,$$

and that there is an exact sequence

$$H_1^{\mathfrak{Fin}}(G;R_{\mathbb{C}}) \rightarrowtail R_{\mathbb{C}}(L) \to R_{\mathbb{C}}(H) \oplus R_{\mathbb{C}}(K) \twoheadrightarrow H_0^{\mathfrak{Fin}}(G;R_{\mathbb{C}})$$

generalizing the previous Corollary.

Exercise

- Show that for $G = \Sigma_3 *_{C_3} \Sigma_3$ one has $H_1^{\mathfrak{Fin}}(G;R_{\mathbb{C}}) \cong \mathbb{Z}$.
- Show that for $G = D_8 *_{C(D_8)} D_8$ one has $H_0^{\mathfrak{Fin}}(G;R_{\mathbb{C}}) \cong \mathbb{Z}^8 \times \mathbb{Z}/2\mathbb{Z}$.

To be able to do more general computations of Bredon homology $H_*^{\mathfrak{Fin}}$ for groups $G = H *_L K$, without assuming H, L and K to be finite, we can make use of the following construction of a model for $\underline{E}G$, which permits to reduce computations involving $\underline{E}G$ to the models for subgroups $\underline{E}H, \underline{E}L$ and $\underline{E}K$. Note that for any subgroup $U < G$ the proper G-space $\mathrm{Ind}_U^G \underline{E}U = G \times_U \underline{E}U$ maps by a unique G-homotopy class to the universal space $\underline{E}G$, extending the U-homotopy equivalence $\underline{E}U \to \underline{E}G$, obtained by viewing $\underline{E}G$ as a model for $\underline{E}U$.

Lemma 3.31. *Let $G = H *_L K$. Then the homotopy push-out P of*

$$\begin{array}{ccc} G \times_L \underline{E}L & \xrightarrow{\mathrm{Ind}_H^G \phi} & G \times_H \underline{E}H \\ {\scriptstyle \mathrm{Ind}_K^G \psi}\downarrow & & \\ G \times_K \underline{E}K & & \end{array}$$

is a model for $\underline{E}G$; the arrows in the diagram are induced by the universal maps $\phi : H \times_L \underline{E}L \to \underline{E}H$ respectively $\psi : K \times_L \underline{E}L \to \underline{E}K$.

Proof. Recall that the homotopy push-out is given by replacing the arrows by inclusions into mapping cylinders, and by forming then the ordinary push-out P of the new diagram. Clearly, the homotopy push-out P is a proper G-CW-complex. It remains to prove that P^U is contractible for any finite subgroup U of G. To see this, we first note that in the above diagram we can replace each of the three spaces $\underline{E}H$, $\underline{E}L$ and $\underline{E}K$ by the same space $\underline{E}G$, viewed as an H, L, K-space respectively, yielding the new diagram

$$\begin{array}{ccc} G \times_L \underline{E}G & \longrightarrow & G \times_H \underline{E}G \\ \downarrow & & \downarrow \\ G \times_K \underline{E}G & \longrightarrow & P. \end{array}$$

For any $U < G$ we can replace the left G-action on spaces $G \times_U \underline{E}G$ by the diagonal G-action (cf. Section 2). Passing to U-fixed points for a finite subgroup of $U < G$ and noting that $\underline{E}G$ is U-contractible, we infer that P^U is homotopy equivalent the U-fixed points of the homotopy push-out $\tilde{P}$ in the diagram

$$\begin{array}{ccc} G/L & \longrightarrow & G/H \\ \downarrow & & \downarrow \\ G/K & \longrightarrow & \tilde{P}. \end{array}$$

But this homotopy push-out $\tilde{P}$ is just the standard construction of a tree on which G acts with vertex stabilizers conjugate to H resp. K and edge stabilizers conjugate to L. Therefore $\tilde{P}^U$ is contractible, being the fixed point space of a finite group acting on a tree (cf. [122]). □

Applied to our case of interest this yields the following.

Corollary 3.32. *Let $G = H *_L K$ with H, K and L arbitrary groups and let $N \in G$-$\mathrm{Mod}_{\mathfrak{Fin}}$. Then there is a long exact Mayer–Vietoris sequence*

$$\begin{array}{ccccc} \cdots \longrightarrow & H_n^{\mathfrak{Fin}}(L;N) & \longrightarrow & H_n^{\mathfrak{Fin}}(H;N) \oplus H_n^{\mathfrak{Fin}}(K;N) \\ & & & \downarrow \\ \cdots \longleftarrow & H_{n-1}^{\mathfrak{Fin}}(L;N) & \xleftarrow{\partial} & H_n^{\mathfrak{Fin}}(G;N) \end{array}$$

Proof. We apply $H_*^{\mathfrak{Fin}}$ to the homotopy-push-out of the previous Lemma 3.31 and obtain a long exact sequence

$$\begin{array}{ccc} \cdots H_n^{\mathfrak{Fin}}(G \times_L \underline{E}L; N) & \longrightarrow & H_n^{\mathfrak{Fin}}(G \times_H \underline{E}H; N) \oplus H_n^{\mathfrak{Fin}}(G \times_K \underline{E}K; N) \\ \longrightarrow H_n^{\mathfrak{Fin}}(\underline{E}G; N) & \xrightarrow{\partial} & H_{n-1}^{\mathfrak{Fin}}(G \times_L \underline{E}L; N) \longrightarrow \cdots \end{array}$$

Since for any U-CW-complex X and subgroup $U < G$ we have an adjunction isomorphism

$$\underline{C_*(G \times_U X)} \otimes_{\mathfrak{Fin}(G)} N \cong \underline{C_*(X)} \otimes_{\mathfrak{Fin}(U)} (N|U)$$

there is an induction isomorphism

$$\mathrm{Ind}_U^G : H_*^{\mathfrak{Fin}(U)}(X; N|U) \cong H_*^{\mathfrak{Fin}(G)}(G \times_U X; N)$$

and the result follows. □

4. Homology and Cohomology Theories

The *spectra* in the topologist's sense are convenient objects to describe (co)homology theories in the category of CW-complexes. We will give a brief outline of some definitions; for details, the reader is referred to Switzer's book [128] or, for a modern approach, to the paper by Elmendorfer, Kriz, Mandell and May [44]. A *spectrum* is a collection of *pointed* CW-complexes $\mathbf{S} = \{S_i \mid i \in \mathbb{N}\}$ together with pointed structure maps $\sigma_i : \Sigma S_i \to S_{i+1}$. Morphisms of spectra $\mathbf{f} : \mathbf{S} \to \mathbf{T}$ are families of pointed maps $f_i : S_i \to T_i$ commuting with the structure maps. Spectra form a category, which we denote by $\mathfrak{SP}$. The *homotopy groups* $\pi_k(\mathbf{S})$ of a spectrum $\mathbf{S}$ are defined by

$$\pi_k(\mathbf{S}) = \mathrm{colim}_i\, \pi_{k+i}(S_i),$$

with the direct limit taken using the structure maps σ_i:

$$\pi_{k+i}(S_i) = [S^{k+i}, S_i]_\bullet \xrightarrow{\mathrm{susp}} [S^{k+i+1}, \Sigma S_i]_\bullet \xrightarrow{\sigma_{i\sharp}} [S^{k+i+1}, S_{i+1}]_\bullet = \pi_{k+i+1}(S_{i+1}).$$

Notice that the groups $\pi_k(\mathbf{S})$ can be non-zero for negative values of k. The smash product of a *pointed* CW-complex Y with a spectrum $\mathbf{S}$ yields a new spectrum $Y \wedge \mathbf{S}$ in an obvious way. One also defines homotopy classes of maps between spectra $\mathbf{S}$ and $\mathbf{T}$, $[\mathbf{S}, \mathbf{T}]$, which form naturally an abelian group such that suspension yields an isomorphism (cf. [128]):

$$[\mathbf{S}, \mathbf{T}] \cong [S^1 \wedge \mathbf{S}, S^1 \wedge \mathbf{T}].$$

The simplest example of a spectrum is the *sphere* spectrum $\mathbf{Sph}$, which has $S_i = S^i$ and $\sigma_i = Id$ (up to a homeomorphism). The resulting groups

$$\pi_k(\mathbf{Sph}) = \pi_k^{st}(S^0)$$

are the *stable* homotopy groups of the zero sphere. More generally, for any pointed space Y one has

$$\pi_k(Y \wedge \mathbf{Sph}) = \pi_k^{st}(Y).$$

These stable homotopy groups define a *reduced (generalized)* homology theory $\tilde{\mathbf{h}}$ on pointed CW-complexes. To define so-called *unreduced* (co)homology theories on arbitrary, not necessarily pointed CW-complexes, one replaces the CW-complex X by the pointed CW-complex $X_+ = X \amalg \{*\}$. Every spectrum $\mathbf{S}$ gives rise to an unreduced generalized (co)homology theory on CW-complexes, and conversely every unreduced generalized (co)homology theory $\mathbf{h}$ on CW-complexes (satisfying the usual axioms, including: *homotopy invariance, Mayer–Vietoris Axiom and the General Disjoint Union Axiom*) is representable by a spectrum in the sense that

$$\mathbf{h}_i(X) := \tilde{\mathbf{h}}_i(X_+) := \pi_i(X_+ \wedge \mathbf{S}) = [S^i \wedge \mathbf{Sph}, X_+ \wedge \mathbf{S}],$$

and

$$\mathbf{h}^i(X) := \tilde{\mathbf{h}}^i(X_+) := [X_+ \wedge \mathbf{Sph}, S^i \wedge \mathbf{S}].$$

To define $\mathbf{h}$ on CW-pairs, one can put

$$\mathbf{h}(X, A) := \tilde{\mathbf{h}}(X_+ \cup CA_+) \cong \tilde{\mathbf{h}}(X/A)$$

where we use the notation CA_+ to denote the cone over A_+ and the convention that $X/A = X_+$ in case A is empty. In particular, $\mathbf{h}(X, \emptyset) = \mathbf{h}(X)$.

If one writes X as a union of subcomplexes X_α such that $X = \operatorname{colim}_\alpha X_\alpha$ then

$$\mathbf{h}_i(X) = \operatorname{colim}_\alpha \mathbf{h}_i(X_\alpha).$$

For instance, if X is written as the union of its finite subcomplexes, this applies. This can be paraphrased by saying that generalized homology theories (given by spectra) have *compact supports.*

If one wishes to extend a generalized (co)homology theory to the category of *all* spaces, one can do this by defining $\mathbf{h}(X)$ as $\mathbf{h}(\Gamma(X))$, where $\Gamma(X)$ denotes the geometric realization functor considered earlier. The resulting (co)homology theory then obviously satisfies the strong form of the homotopy axiom, turning weak homotopy equivalences into isomorphisms.

Every spectrum is equivalent to an Ω-spectrum, that is, a spectrum $\mathbf{S}$ for which the adjoints $S_i \to \Omega S_{i+1}$ of the structure maps σ_i are homotopy equivalences. In case one uses an Ω-spectrum to represent $\mathbf{h}$, the cohomology groups can be expressed as *ordinary* homotopy groups:

$$\mathbf{h}^i(X) = [X_+, S_i]_\bullet = [X, S_i].$$

The most famous examples of Ω-spectra are the Eilenberg–Mac Lane spectrum $\mathbf{H}$ which represents ordinary (co)homology, and the Bott spectrum $\mathbf{BU}$ representing complex K-(co)homology. They are given by $\mathbf{H} = \{K(\mathbb{Z}, i)\}$, with $\Sigma K(\mathbb{Z}, i) \to K(\mathbb{Z}, i+1)$ being the adjoint of the natural equivalence $K(\mathbb{Z}, i) \to \Omega K(\mathbb{Z}, i+1)$, and $\mathbf{BU}=\{BU_i\}$ with $BU_i = \mathbb{Z} \times BU$ for i even, and $BU_i = U$ for i odd; the adjoint of $\Sigma BU_i \to BU_{i+1}$ corresponds to $\mathbb{Z} \times BU \simeq \Omega U$ respectively $U \simeq \Omega(\mathbb{Z} \times BU)$. One writes $K^i(X)$ for the cohomology groups associated to $\mathbf{BU}$, and $K_i(X)$ for the corresponding homology groups; sometimes, these groups are referred to as *representable* K-(co)homology groups. Thus

$$\text{for } i \text{ even:} \quad K_i(X) = \pi_i(X_+ \wedge \mathbf{BU}); \quad K^i(X) = [X, \mathbb{Z} \times BU]$$

and

$$\text{for } i \text{ odd:} \quad K_i(X) = \pi_i(X_+ \wedge \mathbf{BU}); \quad K^i(X) = [X, U].$$

For an arbitrary compact space Z, the group $K^0(Z) := [Z, \mathbb{Z} \times BU]$ agrees with the *Grothendieck group of complex vector bundles over* Z; similarly for $K^1(Z)$. The *K-homology* groups $K_i(Z)$ also admit a geometric interpretation, as certain *bordism* groups (cf. Baum–Douglas [6], see also M. Jakob [65]).

Because $BU \simeq BSU \times K(\mathbb{Z}, 2)$ and $U \simeq SU \times K(\mathbb{Z}, 1)$ with BSU 3-connected and SU 2-connected, one has for a CW-complex X

$$K^0(X) \cong H^0(X;\mathbb{Z}) \oplus H^2(X;\mathbb{Z}), \quad \text{if } \dim X \leq 3$$

and

$$K^1(X) \cong H^1(X;\mathbb{Z}) \cong \prod_{[x_0]\in\pi_0(X)} \mathrm{Hom}(\pi_1(X, x_0), \mathbb{Z}), \quad \text{if } \dim X \leq 2.$$

(One checks that these bijections are actually group isomorphisms. For instance, the homomorphism $K^0(X) \to H^2(X;\mathbb{Z})$ is induced by the first Chern class $c_1 : BU \to K(\mathbb{Z}, 2)$, and the determinant map $U \to U(1) = K(\mathbb{Z}, 1)$ induces the homomorphism $K^1(X) \to H^1(X;\mathbb{Z})$.) To obtain an analogous result for *K-homology* of low dimensional CW-complexes, one can use the universal coefficient sequence relating K-cohomology to K-homology. For an arbitrary CW-complex X there is a natural *universal coefficient sequence* [132]

$$\mathrm{Ext}(K_{i-1}(X), \mathbb{Z}) \rightarrowtail K^i(X) \twoheadrightarrow \mathrm{Hom}(K_i(X), \mathbb{Z}).$$

Comparing it with the universal coefficient sequence for ordinary (co)homology

$$\mathrm{Ext}(H_{i-1}(X;\mathbb{Z}), \mathbb{Z}) \rightarrowtail H^i(X;\mathbb{Z}) \twoheadrightarrow \mathrm{Hom}(H_i(X;\mathbb{Z}), \mathbb{Z}),$$

one concludes that the following holds (first, one should consider the case of a finite X and establish the Lemma below for that case; then one can pass to an arbitrary X by taking a colimit over finite subcomplexes).

Lemma 4.1. *If X is a two dimensional CW-complex, then there are natural isomorphisms*

$$K_0(X) \cong H_0(X;\mathbb{Z}) \oplus H_2(X;\mathbb{Z}); \quad K_1(X) \cong H_1(X;\mathbb{Z}).$$

All spectra are rationally equivalent to products of Eilenberg–Mac Lane spectra, and therefore one can define for an arbitrary homology theory $\mathbf{h}$ given by a spectrum, a *generalized Chern character*

$$Ch_n : \bigoplus_{i+j=n} H_i(X; \mathbf{h}_j(\{*\})) \to \mathbf{h}_n(X) \otimes \mathbb{Q}, \quad n \in \mathbb{Z},$$

which is an isomorphism upon tensoring with $\mathbb{Q}$. The map Ch_n is determined as a natural transformation of homology theories by its values on spheres. An explicit construction which we will sketch here, is described in Dold's note [37]. If $\mathbf{h}$ is given by the spectrum $\mathbf{S}$ then there is a *Hurewicz pairing*

$$\psi : \pi_i^{st}(X_+) \otimes \mathbf{h}_*(\{pt\}) \longrightarrow \mathbf{h}_{i+*}(X)$$

which is defined by mapping $f \otimes \alpha$ with $f \in \pi_i^{st}(X_+)$ and $\alpha \in \mathbf{h}_*(\{pt\})$, to $f_*(\alpha)$, where f_* is the map induced by f via

$$f_* : \mathbf{h}_*(\{pt\}) = \tilde{\mathbf{h}}_*(S^0) \to \tilde{\mathbf{h}}_{*+i}(X_+) = \mathbf{h}_{*+i}(X).$$

In case of ordinary homology, $H_*(\{pt\};\mathbb{Z}) \cong \mathbb{Z}$ with a distinguished generator ι, the Hurewicz pairing can be replaced by

$$\mathrm{Hu} : \pi_i^{st}(X_+) \to H_i(X;\mathbb{Z}), \quad f \mapsto f_*(\iota),$$

the classical Hurewicz Homomorphism. Note that $\mathrm{Hu} \otimes \mathbb{Q}$ is an isomorphism, since it is an isomorphism on spheres (the higher stable homotopy groups $\pi_k^{st}(S^n)$, $k > n$ are all finite).

The Chern map Ch_n can now be defined by describing its components as follows

$$\begin{array}{ccc} H_i(X;\mathbf{h}_j(\{*\})) & \xrightarrow{-\otimes\mathbb{Q}} & H_i(X;\mathbb{Q}) \otimes \mathbf{h}_j(\{*\}) \\ & & \cong \big\downarrow (\mathrm{Hu} \otimes h_j(\{*\}) \otimes \mathbb{Q})^{-1} \\ & & \pi_i^{st}(X_+) \otimes \mathbf{h}_j(\{*\}) \otimes \mathbb{Q} \xrightarrow{\psi \otimes \mathbb{Q}} \mathbf{h}_n(X) \otimes \mathbb{Q}, \end{array}$$

with ψ the Hurewicz pairing as defined above. This translates to the following.

Lemma 4.2. *For an arbitrary CW-complex X one has natural isomorphisms*

$$K_0(X) \otimes \mathbb{Q} \cong \bigoplus_i H_{2i}(X;\mathbb{Q}); \quad K_1(X) \otimes \mathbb{Q} \cong \bigoplus_i H_{2i+1}(X;\mathbb{Q}).$$

The previous two lemmas are special cases of the following. It is well-known that the classifying space BU has k-invariants $\{k^j\}$ satisfying $k^{2i} = 0$ and k^{2i+1} of order $(i-1)!$, see [108]. Using this fact one can show that if X is a CW-complex of dimension $\leq 2N$, the Chern character can already be defined after inverting $N!$, giving rise to

$$(K_0(X) \oplus K_1(X)) \otimes \mathbb{Z}[1/N!] \cong \bigoplus_i H_i(X;\mathbb{Z}[1/N!]).$$

Remark 4.3. Dold's Chern map yields in the case of K-homology a natural transformation of homology theories

$$Ch_* : \bigoplus_{i \in \mathbb{Z}} (H_{2i}(X;\mathbb{Z}) \oplus H_{2i+1}(X;\mathbb{Z})) \to (K_0(X) \oplus K_1(X)) \otimes \mathbb{Q}$$

which, for $X = S^n$, maps $H_*(S^n;\mathbb{Z})$ isomorphically onto the image of

$$K_*(S^n) \to K_*(S^n) \otimes \mathbb{Q}.$$

From this, one can conclude that the constructed Chern map ch_* is, in even degrees, dual to the classical Chern character

$$ch : K^0(X) \to \prod H^{2k}(X;\mathbb{Q}),$$

which for a finite complex X and vector bundle ξ over X is given by $ch(\xi) = \sum s_k(\xi)/k!$, where s_k denotes the k-th Newton polynomial in the Chern classes of ξ.

Remark 4.4. It is useful to know that one has for a general CW-complex a natural injection (*Hurewicz Homomorphism*)

$$\mathrm{Hu}_1 : H_1(X;\mathbb{Z}) \rightarrowtail K_1(X).$$

To define this map we may assume X to be pointed and connected. Choose a generator λ of $K_1(S^1) \cong \mathbb{Z}$ and use the usual Hurewicz isomorphism $H_1(X;\mathbb{Z}) \cong \pi_1(X)_{ab}$ to define Hu_1 by

$$H_1(X;\mathbb{Z}) \xleftarrow{\cong} \pi_1(X)_{ab} \ni [f] \mapsto f_*(\lambda) \in K_1(X).$$

To show that Hu_1 is injective, we may assume that X is a finite, pointed and connected complex. It is clear from our rational discussions above that $\mathrm{Hu}_1 \otimes \mathbb{Q}$ is injective, so we will concentrate on torsion elements in the finitely generated abelian group $H_1(X;\mathbb{Z})$. If there is an element of finite order in the kernel, then there is also a finite cyclic direct summand $S \subset H_1(X;\mathbb{Z})$ on which Hu_1 is not injective. Suppose S is cyclic of order n. Then there exist a map $X \to K(\mathbb{Z}/n\mathbb{Z}, 1)$, corresponding to a map $\pi_1(X) \to \mathbb{Z}/n\mathbb{Z}$, such that the induced map in integral homology maps S isomorphically onto $H_1(K(\mathbb{Z}/n\mathbb{Z},1);\mathbb{Z})$. It therefore suffices to check that Hu_1 is injective for the case of $X = K(\mathbb{Z}/n\mathbb{Z},1)$. For this, one can consider the Atiyah–Hirzebruch spectral sequence [128]

$$\bigoplus_{i+j=m} H_i(K(\mathbb{Z}/n\mathbb{Z},1);K_j(\{*\})) \Rightarrow K_m(K(\mathbb{Z}/n\mathbb{Z},1))$$

which collapses, because for an even $i > 0$ one has $H_i(K(\mathbb{Z}/n\mathbb{Z},1);\mathbb{Z}) = 0$. The resulting edge homomorphism

$$\mathrm{Hu}_1 : H_1(K(\mathbb{Z}/n\mathbb{Z},1);\mathbb{Z}) \to K_1(K(\mathbb{Z}/n\mathbb{Z},1))$$

is therefore injective. (One can use this edge homomorphism as the *definition* of Hu_1, if one wants to avoid identifying the former with the latter.)

5. Spaces and Spectra over an Orbit Category

The main reference for this section is the seminal paper by J. F. Davis and W. Lück [31]. Let G be a group and $\mathfrak{F}$ a family of subgroups of G. We will assume in this section that $\mathfrak{F}$ is non-empty and closed under conjugation and passing to subgroups. We write $\mathfrak{CW}$ for the category of CW-complexes and, as earlier, $\mathfrak{SP}$ for the category of spectra.

An $\mathfrak{D}_{\mathfrak{F}}(G)$-space is a *contravariant* functor

$$\mathfrak{D}_{\mathfrak{F}}(G) \longrightarrow \mathfrak{CW}.$$

A typical example of an $\mathfrak{O}_{\mathfrak{F}}(G)$-space is given by considering a G-CW-complex X as an $\mathfrak{O}_{\mathfrak{F}}(G)$-space, also denoted by X, with $X(G/H) := X^H$ and for $f : G/K \to G/H$ given by $f(K) = gH$, $X(f) = f^* : X^H \to X^K$ the map $x \mapsto gx$.

An $\mathfrak{O}_{\mathfrak{F}}(G)$-spectrum $\mathbf{S}$ is a *covariant* functor

$$\mathbf{S} : \mathfrak{O}_{\mathfrak{F}}(G) \to \mathfrak{SP}.$$

Such an $\mathfrak{O}_{\mathfrak{F}}(G)$-spectrum $\mathbf{S}$ gives rise to a G-homology theory as follows. If X is a G-CW-complex one defines the spectrum $X_+ \otimes_G \mathbf{S}$ by

$$X_+ \otimes_G \mathbf{S} = \coprod_{G/H \in \mathfrak{O}_{\mathfrak{F}}(G)} (X_+^H \wedge \mathbf{S}(G/H))/\sim,$$

where the equivalence relation is generated by $f^*x \wedge y \sim x \wedge f_* y$, with $f : G/K \to G/H$, $x \in X_+^H$, $y \in S_i(G/K)$ for some i (the morphism f_* is $\mathbf{S}(f)$). One then puts

$$\mathbf{h}_i^G(X) = \pi_i(X_+ \otimes_G \mathbf{S}), \quad i \in \mathbb{Z},$$

to define the value of the associated G-homology theory $\mathbf{h}_*^G$ on G-spaces X; similarly, one defines relative groups $\mathbf{h}_*^G(X, A)$. Such a G-homology theory satisfies the analogues of the Eilenberg–Steenrod axioms for G-CW-complexes. For instance, the *disjoint union axiom* states that if the G-CW-complex X is the disjoint union of an arbitrary family $\{X_\alpha\}$ of G-CW-complexes, then there is a natural isomorphism

$$\mathbf{h}_*^G(X) \cong \bigoplus \mathbf{h}_*^G(X_\alpha).$$

Also, the long exact sequence of the pair $S^0 \times X \subset B^1 \times X$ with trivial G-action on B^1, yields the natural *suspension isomorphism*

$$\sigma_i : \mathbf{h}_i^G((B^1, S^0) \times X) \xrightarrow{\cong} \mathbf{h}_{i-1}^G(X).$$

Remark 5.1. One could also consider *covariant* $\mathfrak{O}_{\mathfrak{F}}(G)$-spaces and *contravariant* $\mathfrak{O}_{\mathfrak{F}}(G)$-spectra (cf. [31]), but we won't need these here; the latter are relevant in the context of G-*cohomology* theories.

If $\mathfrak{F}$ is the family of *all* subgroups of G, we write $\mathfrak{O}_{\mathfrak{All}}$ or just $\mathfrak{O}(G)$ instead of $\mathfrak{O}_{\mathfrak{F}}(G)$. For $H < G$ a subgroup and $\mathbf{S}$ an $\mathfrak{O}(G)$-spectrum, one has an obvious notion of restriction to obtain an $\mathfrak{O}(H)$-spectrum $\mathbf{S}|H$. On the other hand, an H-CW-complex X gives rise to an induced $\mathfrak{O}(G)$-space $\mathrm{Ind}_H^G X =: G \times_H X$, $G/L \mapsto (G \times_H X)^L$. There is an adjunction equivalence of spectra

$$X_+ \otimes_H (\mathbf{S}|H) \simeq (G \times_H X)_+ \otimes_G \mathbf{S}.$$

Corollary 5.2. *Let $\mathbf{S}$ be an $\mathfrak{O}_{\mathfrak{All}}(G)$-spectrum with associated homology theory $\mathbf{h}_*^G$ and let $H < G$ be a subgroup. Denote by $\mathbf{h}_*^H$ the homology theory defined by $\mathbf{S}|H$. Then for X an H-CW-complex there is a natural induction isomorphism*

$$\mathbf{h}_i^H(X) \cong \mathbf{h}_i^G(G \times_H X), \quad i \in \mathbb{Z}.$$

If G is the directed union of subgroups G_α, then one has a natural isomorphism

$$\mathbf{h}_i^G(\underline{E}G) \cong \operatorname{colim}_\alpha \mathbf{h}_i^{G_\alpha}(\underline{E}G_\alpha).$$

Proof. The first statement follows from the adjunction equivalence mentioned above. For the second statement one uses the functorial model for $\underline{E}G$ described in Section 2, which is the directed union of G-subcomplexes of the form $G \times_{G_\alpha} \underline{E}G_\alpha$. The direct limit axiom for representable equivariant homology theories then implies that

$$\mathbf{h}_i^G(\underline{E}G) \cong \operatorname{colim}_\alpha \mathbf{h}_i^G(G \times_{G_\alpha} \underline{E}G_\alpha),$$

and the result follows by applying the adjunction equivalences of the first part, with $H = G_\alpha$. □

Recall that for a (discrete) group G the reduced C^*-algebra $C_r^*(G)$ is defined to be the norm closure of the complex group algebra $\mathbb{C}[G]$ viewed as a subalgebra of the Banach algebra $\mathcal{B}(\ell_2(G))$ of bounded operators $\ell_2(G) \to \ell_2(G)$; $\ell_2(G)$ stands for the Hilbert space of square summable functions $f : G \to \mathbb{C}$, which we also denote by $f = \sum_G f(g) \cdot g$ with $\sum_G |f(g)|^2 < \infty$. One then defines *topological* algebraic K-groups $K_i^{top}(C_r^*(G))$, $i \in \mathbb{Z}$, which satisfy Bott periodicity [131] (see also Karoubi's book [69]; many basic facts on K-theory of Banach algebras can also be found in Rosenberg's book [113]). In particular, $K_0^{top}(C_r^*(G)) = K_0^{alg}(C_r^*(G))$ is the projective class group of the ring $C_r^*(G)$; we sometimes just write $K_0(C_r^*(G))$ for this group, since the topology of $C_r^*(G)$ is here irrelevant. By definition, $K_1^{top}(C_r^*(G)) = \pi_0(\operatorname{colim}_n Gl_n(C_r^*(G)))$, where $Gl_n(C_r^*(G))$ is given the subspace topology from the inclusion $Gl_n(C_r^*(G)) \subset (C_r^*(G))^{n^2}$. Later on, we will also consider the so-called *maximal* C^*-algebra $C^*(G)$ of G, which is the completion of $\mathbb{C}[G]$ with respect to the norm given by the supremum over all norms coming from irreducible unitary representations of $\mathbb{C}[G]$ on some Hilbert space. Many technical results on the K-theory of C^*-algebras require these Banach algebras to be separable; of course, this condition is for $C_r^*(G)$ and $C^*(G)$ only satisfied if G is countable. We recall that a group G is the *directed union* of the subgroups $\{G_\alpha\}_{\alpha \in I}$ if G is the union of the groups G_α and I is a partially ordered set such that the following two properties hold: first, for indices $\alpha \leq \beta$ in I one has $G_\alpha \subset G_\beta$, and second, for each pair α, β in I there exists a $\gamma \in I$ satisfying $\alpha \leq \gamma$ an $\beta \leq \gamma$. In particular, G is naturally isomorphic to the colimit of the groups G_α with respect to I. Clearly, any group is the directed union of its countable (or finitely generated) subgroups. It is useful to know the following results.

Lemma 5.3. *Let G be the directed union of subgroups $\{G_\alpha\}$. Then one has natural isomorphisms*

$$K_*^{top}(C_r^*(G)) \cong \operatorname{colim}_\alpha K_*^{top}(C_r^*(G_\alpha))$$

and

$$K_*^{top}(C^*(G)) \cong \operatorname{colim}_\alpha K_*^{top}(C^*(G_\alpha)).$$

Proof. The maps $C_r^*(G_\alpha) \to C_r^*(G)$ are injective morphisms of C^*-algebras, thus isometric embeddings. As a result, $C_r^*(G)$ contains the algebraic direct limit of the algebras $C_r^*(G_\alpha)$ as a dense subalgebra. The result concerning their K-theory then follows by standard C^*-algebra techniques (cf. Higson and Roe [63], Remark 4.1.18). Similarly one deals with the case of the maximal C^*-algebras. □

The following are some basic examples of reduced C^*-algebras of groups. The simplest one is that of a finite group. If H is finite, $\mathbb{C}[H] = C_r^*(H)$, and $K_0(\mathbb{C}[H]) = R_\mathbb{C}(H)$, the complex representation ring of H, and $K_1^{top}(\mathbb{C}[H]) = 0$, because $Gl_n(\mathbb{C}[H])$ is connected. Indeed, as a topological group, $Gl_n(\mathbb{C}[H])$ is isomorphic to a finite product $\prod Gl_{n_k}(\mathbb{C})$, because $\mathbb{C}[H]$ is a product of complex matrix rings, and the groups $Gl_{n_k}(\mathbb{C})$ are connected so that $\pi_0(Gl_\infty(\mathbb{C}[H])) = 0$, with $Gl_\infty(\mathbb{C}[H]) = \operatorname{colim}_n Gl_n(\mathbb{C}[H])$. On the other hand, the *algebraic* K-group

$$K_1^{alg}(\mathbb{C}[H]) := Gl_\infty(\mathbb{C}[H])/[Gl_\infty(\mathbb{C}[H]), Gl_\infty(\mathbb{C}[H])]$$

has the form $\prod \mathbb{C}^\times$ and is non-zero, even for H the trivial group (we use here the well-known fact that $K_1^{alg}(M_n(\mathbb{C}))$ is isomorphic to $K_1^{alg}(\mathbb{C})$ and the latter is isomorphic to $\mathbb{C}^\times$ by the determinant map $Gl_\infty(\mathbb{C}) \to \mathbb{C}^\times$).

Another important example is that of $C_r^*(\mathbb{Z})$. By identifying $\ell_2(\mathbb{Z})$ with the Hilbert space $L_2(S^1)$ of square integrable complex valued functions on the circle, one can show that $C_r^*(\mathbb{Z})$ is isomorphic as a C^*-algebra to the C^*-algebra $C(S^1)$ of complex valued *continuous* functions on S^1; the action of $C(S^1)$ on $L_2(S^1)$ is the obvious one (point-wise multiplication). From the theory of vector bundles it is well-known that for any compact space X, $K_i^{top}(C(X))$ is naturally isomorphic to the topological K-cohomology group $K^i(X)$, defined by the Bott-spectrum **BU**. It follows that

$$K_i^{top}(C_r^*(\mathbb{Z})) \cong K^i(S^1) \cong \mathbb{Z} \quad \text{for all } \ i \in \mathbb{Z},$$

and, by a similar argument,

$$K_i^{top}(C_r^*(\mathbb{Z}^k)) \cong \mathbb{Z}^{2^{k-1}}.$$

The representation on $\ell_2(G)$ considered above is referred to as the *regular* representation of G. Thus, there is a natural continuous surjection $C^*(G) \twoheadrightarrow C_r^*(G)$, which is known to be an isomorphism if and only if the group G is amenable [107]; for instance $C^*(\mathbb{Z}) \cong C_r^*(\mathbb{Z})$. For the convenience of the reader we recall one of many equivalent definitions of amenable (discrete) groups.

Definition 5.4. Let G be a discrete group and let $\ell_\infty(G, \mathbb{R})$ denote the Banach algebra of bounded functions $\phi : G \to \mathbb{R}$, with left G-action on $\ell_\infty(G, \mathbb{R})$ being defined by $(g\phi)(x) = \phi(xg)$. Then G is called *amenable* if there exists a bounded linear functional $M : \ell_\infty(G, \mathbb{R}) \to \mathbb{R}$, such that

- $M(g\phi) = M(\phi), \quad \forall g \in G$
- $M(\text{constant function } 1) = 1$
- $\phi \geq 0 \Longrightarrow M(\phi) \geq 0.$

The map M is called a *mean.* Finite groups are obviously amenable. Solvable groups are also known to be amenable. The class of amenable groups is closed under extensions, passing to subgroups and factor groups, and also with respect to directed unions; the free group on two generators is not amenable. Thus, an amenable group never contains a non-abelian free subgroup. There is a related notion of *K-amenability* due to Cuntz. If G is K-amenable, then the natural map $K_*^{top}(C^*(G)) \to K_*^{top}(C_r^*(G))$ is an isomorphism [30]. For instance, countable free groups and countable amenable groups are K-amenable. The class of K-amenable groups is closed under finite products and passing to subgroups [30]. For our purpose it is useful to consider the following slightly weaker notion.

Definition 5.5. A group G is called *K_*-amenable*, if all its countable subgroups are K-amenable.

Note that subgroups of K_*-amenable groups are K_*-amenable too, because subgroups of K-amenable groups are K-amenable. Moreover, it then follows from the definition, that also a directed union of K_*-amenable groups is K_*-amenable. In particular, all free groups and all amenable groups are K_*-amenable. Also, finite direct products of K_*-amenable groups are K_*-amenable. Note that in view of Lemma 5.3, for all K_*-amenable groups G the natural map

$$K_*^{top}(C^*(G)) \to K_*^{top}(C_r^*(G))$$

is an isomorphism. (Examples of groups G for which $K_0^{top}(C^*(G)) \to K_0^{top}(C_r^*(G))$ is not an isomorphism include the groups $Sl(n, \mathbb{Z})$ for $n \geq 3$; thus, these groups are not K_*-amenable.)

If a countable group G is the fundamental group of a countable graph of K-amenable groups then G is K-amenable (cf. Pimsner [109]). This can now easily be generalized to not necessarily countable groups as follows.

Corollary 5.6. *Let G be the fundamental group of a graph of K_*-amenable groups. Then G is K_*-amenable and therefore the natural map*

$$K_*^{top}(C^*(G)) \to K_*^{top}(C_r^*(G))$$

is an isomorphism.

Proof. Suppose G is the fundamental group of a graph of groups $\{G_\alpha\}$ each of which is K_*-amenable. By the theory of graphs of groups, G is the directed union of subgroups $\{H_\beta\}$ with each H_β the fundamental group of a countable graph of countable groups $\{H_{\beta,\gamma}\}$ in such a way that each $H_{\beta,\gamma}$ lies in some G_α and, therefore, is K_*-amenable. But since the groups $\{H_{\beta,\gamma}\}$ are countable, they are actually K-amenable. Now Pimsner's result mentioned earlier implies that each H_β is K-amenable. But this implies that G is K_*-amenable, being the directed union of the K_*-amenable groups $\{H_\beta\}$. □

As we will see, the *Baum–Connes Conjecture* relates the K-theory of $C_r^*(G)$ to the representation rings of the finite subgroups of G via an equivariant version of K-homology of the G-space $E\mathfrak{Fin}(G)=\underline{E}G$. This equivariant version of K-homology is defined by means of the *non-connective topological K-theory spectrum* $\mathbf{K}^{top,G}$ constructed in [31] (we will denote it by $\mathbf{K}^{top}$, when G is clear from the context). It is an $\mathfrak{O}(G)$-spectrum with the property that

$$\pi_*(\mathbf{K}^{top}(G/H)) \cong K_*^{top}(C_r^*(H)).$$

For a G-CW-complex X we put

$$K_i^G(X) := \pi_i(X_+ \otimes_G \mathbf{K}^{top}), \quad i \in \mathbb{Z},$$

so that K_*^G is the G-homology theory associated to the $\mathfrak{O}(G)$-spectrum $\mathbf{K}^{top,G}$. It follows that

$$K_i^G(\{*\}) \cong K_i^{top}(C_r^*(G)).$$

By construction of $\mathbf{K}^{top,G}$ (cf. [31]) for $H < G$ a subgroup one has $\mathbf{K}^{top,G}|H \simeq \mathbf{K}^{top,H}$, and therefore equivariant K-homology comes with a natural *induction structure.* Corollary 5.2 now implies the following.

Corollary 5.7. *Let $H < G$ and let X be an H-CW-complex. Then there is a natural isomorphism*

$$K_i^H(X) \xrightarrow{\cong} K_i^G(G \times_H X), \quad i \in \mathbb{Z}.$$

Moreover, if G is the directed union of subgroups $\{G_\alpha\}$, then one has a natural isomorphism

$$K_i^G(\underline{E}G) \cong \operatorname{colim}_\alpha K_i^{G_\alpha}(\underline{E}G_\alpha), \quad i \in \mathbb{Z}.$$

The construction of $\mathbf{K}^{top,G}$ is such that the trivial group yields $K_*^{\{e\}} = K_*$, ordinary K-homology given by the Bott-spectrum, so that for Y an arbitrary CW-complex one has a natural isomorphism

$$K_i(Y) \cong K_i^G(G \times Y).$$

On the other hand, if X is a free G-CW-complex the general definition of G-homology groups using an $\mathfrak{O}_{\mathfrak{F}}(G)$-spectrum $\mathbf{S}$ shows that $X_+ \otimes_G \mathbf{S}$ reduces to $(X/G)_+ \otimes \mathbf{S}|\{e\}$ so that

$$K_i^G(X) \cong K_i(X/G), \quad X \text{ a free } G\text{-}CW\text{-complex}.$$

More generally, if X is a G-CW-complex and the normal subgroup $H < G$ acts freely on X, then $K_i^G(X) \cong K_i^{G/H}(X/H)$. Combined with Corollary 5.7 one sees that the following holds.

Corollary 5.8. *Let X be a G-CW-complex and $\phi : G \to K$ a homomorphism such that $\ker\phi$ acts freely on X. Then one has a natural induction isomorphism*

$$\operatorname{Ind}_* : K_*^G(X) \to K_*^K(\operatorname{Ind}_\phi X) = K_*^K(K \times_\phi X).$$

Indeed, the map Ind_* is the composition of isomorphisms

$$K_*^G(X) \to K_*^{G/\ker\phi}(X/\ker\phi) \to K_*^K(\mathrm{Ind}_{G/\ker\phi}^K(X/\ker\phi)).$$

An important special case is the case of an inclusion $\iota : G \subset K$. When applied to the case of $\underline{E}G$, there is a natural K-map $K \times_G \underline{E}G \to \underline{E}K$ so that we obtain a natural induced map

$$\iota_* : K_*^G(\underline{E}G) \cong K_*^K(K \times_G \underline{E}G) \to K_*^K(\underline{E}K).$$

It is also possible to define, more generally, for any $\phi : G \to K$ a natural map

$$\phi_* : K_*^G(\underline{E}G) \to K_*^K(\underline{E}K),$$

see Remark 5.11 for more details.

Note also that $K_i^G(EG) \cong K_i(BG)$, where $BG := (EG)/G$ so that the canonical G-map $EG \to \underline{E}G$ gives rise to a natural map

$$K_i(BG) \longrightarrow K_i^G(\underline{E}G),$$

which is an isomorphism in case G is torsion-free, because then $\underline{E}G = EG$. The discrete G-space G/H yields

$$K_i^G(G/H) \cong K_i^H(\{*\}) \cong K_i^{top}(C_r^*(H)).$$

The so called *assembly map* in the Baum–Connes Conjecture is now constructed as follows.

Conjecture 5.9. (Baum–Connes): *The assembly map, induced by the G-map $\underline{E}G \to \{*\}$,*

$$\Theta(G) : K_*^G(\underline{E}G) \longrightarrow K_*^G(\{*\}) = K_*^G(G/G) = K_*^{top}(C_r^*(G)),$$

is an isomorphism.

If X is a cocompact proper G-CW-complex and G is a countable group, the *Kasparov* KK-groups $KK_*^G(C_0(X), \mathbb{C})$ are defined [70] ($C_0(X)$ stands for the C^*-algebra of continuous $\mathbb{C}$-valued functions on X which vanish at infinity). For such X, these groups are known to agree with the groups $K_*^G(X)$ defined above. The original Baum–Connes Conjecture is formulated within the framework of the KK-groups, with an assembly map defined using an equivariant index (cf. [5]); we refer to this assembly map as the *analytic* assembly map. That the analytic *Baum–Connes* assembly map can (for countable G) naturally be identified with the *Davis–Lück* assembly map $\Theta(G)$ of 5.9 is proved in [54]. Of course, if one extends the definition of the KK-groups to arbitrary proper G-CW-complexes Y (as this is usually done) by putting

$$KK_*^G(Y) := \mathrm{colim}_{\{X \subset Y,\ X\ G-\mathrm{cocompact}\}} KK_*^G(C_0(X), \mathbb{C})$$

then $KK_*^G(Y) \cong K_*^G(Y)$. But we emphasize that in the definition of the Davis–Lück assembly map $\Theta(G)$ no countability assumption on G is required.

It is often possible to restrict the attention to the case of countable groups or even finitely generated ones, when dealing with the Baum–Connes Conjecture 5.9. Namely, the following holds.

Theorem 5.10. *Let G be a directed union of subgroups $\{G_\alpha\}$. If each G_α satisfies the Baum–Connes Conjecture 5.9, then so does G.*

Proof. In view of Corollary 5.2 and Lemma 5.3 the assembly map

$$\Theta(G) : K_*^G(\underline{E}G) \to K_*^{top}(C_r^*(G))$$

is the colimit of the assembly maps $\Theta(G_\alpha)$, which by assumption are isomorphisms; the result then follows. □

Remark 5.11. Similarly to the spectrum $\mathbf{K}^{top,G}$ one can construct an $\mathfrak{O}(G)$-spectrum $\mathbf{K}^{top,G}_{max}$ with coefficients involving the *maximal* C^*-algebra rather than the reduced one:

$$\pi_i((G/H)_+ \otimes_G \mathbf{K}^{top,G}_{max}) = K_i^{top}(C^*(H)).$$

The resulting equivariant homology theory has properties analogous to those of K_*^G and agrees with that theory on proper G-CW-complexes, since for finite groups H one has $C^*(H) = C_r^*(H) = \mathbb{C}[H]$. Also, there is a natural transformation of equivariant homology theories which corresponds on coefficients to the natural map of C^*-algebras $C^*(H) \to C_r^*(H)$. The advantage of working with the maximal C^*-algebra is that it is functorial in the group variable (it is easy to see that the reduced one is functorial on the category of groups, with injective homomorphisms as morphisms). Since on proper G-CW-complexes the associated homology theories assume the same values, we can deduce from an *arbitrary* homomorphism $\phi : G \to K$, using the induced morphism $\mathbf{K}^{top,G}_{max} \to \mathbf{K}^{top,K}_{max}$, a map

$$\underline{E}G_+ \otimes_G \mathbf{K}^{top,G}_{max} \to \mathrm{Ind}_\phi \underline{E}G_+ \otimes_K \mathbf{K}^{top,K}_{max} \to \underline{E}K_+ \otimes_K \mathbf{K}^{top.K}_{max},$$

giving rise to

$$\phi_* : K_*^G(\underline{E}G) \to K_*^K(\mathrm{Ind}_\phi \underline{E}G) \to K_*^K(\underline{E}K).$$

Also, the map $\underline{E}G_+ \otimes_G \mathbf{K}^{top,G}_{max} \to \{*\}_+ \otimes_G \mathbf{K}^{top,G}_{max}$ yields an assembly map

$$\Theta_{max}(G) : K_*^G(\underline{E}G) \to K_*^{top}(C^*(G))$$

fitting into a commutative diagram

$$\begin{array}{ccc} K_*^G(\underline{E}G) & \xrightarrow{\phi_*} & K_*^K(\underline{E}K) \\ \Theta_{max}(G)\downarrow & & \downarrow\Theta_{max}(K) \\ K_*^{top}(C^*(G)) & \xrightarrow{\phi_*} & K_*^{top}(C^*(K)). \end{array}$$

Note also that the Baum–Connes assembly map factors as

$$\Theta(G) = \pi_* \circ \Theta_{max}(G) : K_*^G(\underline{E}G) \to K_*^{top}(C^*(G)) \to K_*^{top}(C_r^*(G)),$$

where π_* is induced by the natural projection $C^*(G) \to C_r^*(G)$. However, note that there is no natural induced map $\phi_* : C_r^*(G) \to C_r^*(K)$ associated with an *arbitrary* homomorphism $\phi : G \to K$: for instance, the C^*-algebra $C_r^*(\mathbb{Z} * \mathbb{Z})$ is known to be simple (cf. Powers [110]) so that the homomorphism $\mathbb{Z} * \mathbb{Z} \to \{e\}$ cannot induce a $\mathbb{C}$-algebra map $C_r^*(\mathbb{Z} * \mathbb{Z}) \to \mathbb{C} = \mathbb{C}[\{e\}]$.

The Baum–Connes Conjecture 5.9 will be denoted by **BCC**. We will now discuss some computations of $K_*^G(\underline{E}G)$ and we will show how to prove the **BCC** in some cases. The case of a finite group is easy: the map $\underline{E}G \to \{*\}$ is a G-homotopy equivalence and therefore $\Theta(G)$ is certainly an isomorphism. To prove **BCC** for more general groups, one can use Mayer–Vietoris sequences as follows.

Theorem 5.12. *Suppose that X is a 1-dimensional contractible G-CW-complex. Then there is a natural long exact sequence*

$$\begin{array}{ccccc} \bigoplus_v K_0^{G_v}(\underline{E}G_v) & \longrightarrow & K_0^G(\underline{E}G) & \longrightarrow & \bigoplus_e K_1^{G_e}(\underline{E}G_e) \\ \uparrow & & & & \downarrow \\ \bigoplus_e K_0^{G_e}(\underline{E}G_e) & \longleftarrow & K_1^G(\underline{E}G) & \longleftarrow & \bigoplus_v K_1^{G_v}(\underline{E}G_v). \end{array}$$

The sums are taken over orbits of vertices resp. edges; the maps come from a natural Mayer–Vietoris sequence, as explained in the proof below.

Proof. The G-space X is given as a push-out

$$\begin{array}{ccc} S^0 \times \Delta_1 & \longrightarrow & X^0 \\ \downarrow & & \downarrow \\ B^1 \times \Delta_1 & \longrightarrow & X. \end{array}$$

The discrete G-space $X^0 = \Delta_0$ is of the form $\coprod_v G/G_v$ where v runs over the orbits of vertices of X; similarly, Δ_1 has the form $\coprod_e G/G_e$. Since X is a tree, the fixed point space X^U is contractible for every finite subgroup $U < G$ (cf. Serre's book [122]). It follows that the G-space $X \times \underline{E}G$ is G-homotopy equivalent to $\underline{E}G$ so that, because we can use $\underline{E}G$ also as a model for $\underline{E}G_e$ and $\underline{E}G_v$ by restricting the G-action to G_e respectively G_v, one has a G-homotopy push-out of the form

$$\begin{array}{ccc} S^0 \times (\coprod_e (G \times_{G_e} \underline{E}G_e)) & \longrightarrow & \coprod_v (G \times_{G_v} \underline{E}G_v) \\ \downarrow & & \downarrow \\ B^1 \times (\coprod_e (G \times_{G_e} \underline{E}G_e)) & \longrightarrow & \underline{E}G. \end{array}$$

Therefore, the K^G-theory Mayer–Vietoris sequence of the G-pair

$$(\underline{E}G, \coprod_v (G \times_{G_v} \underline{E}G_v)) = (B^1, S^0) \times (\coprod_e (G \times_{G_e} \underline{E}G_e)),$$

using the suspension isomorphism (page 32)

$$\begin{aligned} K_i^G((B^1,S^0) \times (\coprod_e (G \times_{G_e} \underline{E}G_e))) &\cong K_{i-1}^G(\coprod_e (G \times_{G_e} \underline{E}G_e)) \\ &\cong \bigoplus_e K_{i-1}^G(G \times_{G_e} \underline{E}G_e) \end{aligned}$$

will take the form:

$$\begin{array}{ccccc} \bigoplus_v K_0^G(G \times_{G_v} \underline{E}G_v) & \longrightarrow & K_0^G(\underline{E}G) & \longrightarrow & \bigoplus_e K_1^G(G \times_{G_e} \underline{E}G_e) \\ \uparrow & & & & \downarrow \\ \bigoplus_e K_0^G(G \times_{G_e} \underline{E}G_e) & \longleftarrow & K_1^G(\underline{E}G) & \longleftarrow & \bigoplus_v K_1^G(G \times_{G_v} \underline{E}G_v) \end{array}$$

The proof is now completed by applying the induction isomorphisms

$$K_i^G(G \times_{G_v} \underline{E}G_v) \cong K_i^{G_v}(\underline{E}G_v),$$

and similarly for the terms involving G_e. □

If in the theorem above we apply term-wise the assembly map (i.e., we map the spaces $\underline{E}G_v$, $\underline{E}G_e$ and $\underline{E}G$ to a point), we obtain a map Φ to a sequence of the form

$$\begin{array}{ccccc} \bigoplus_v K_0^{top}(C_r^*(G_v)) & \xrightarrow{\beta_0} & K_0^{top}(C_r^*(G)) & \xrightarrow{\partial_0} & \bigoplus_e K_1^{top}(C_r^*(G_e)) \\ \alpha_0 \uparrow & & & & \downarrow \alpha_1 \\ \bigoplus_e K_0^{top}(C_r^*(G_e)) & \xleftarrow{\partial_1} & K_1^{top}(C_r^*(G)) & \xleftarrow{\beta_1} & \bigoplus_v K_1^{top}(C_r^*(G_v)) \end{array}$$

The maps α_i and β_i are the obvious ones; with suitably defined maps ∂_i, this sequence corresponds to the exact sequence of Pimsner [109], if we assume the group G in question to be countable. According to Oyono-Oyono,[105] still assuming G to be countable, the map of exact sequences Φ is such that all resulting squares are commutative (of course, only the squares involving the maps ∂_i are in question). As a result the following basic theorem, which in the countable case is due to Oyono-Oyono, holds (see also Tu [129] for a closely related result).

Theorem 5.13. *If there exists a 1-dimensional contractible G-CW-complex with all stabilizers satisfying the Baum–Connes Conjecture* **BCC**, *then G satisfies* **BCC**.

As a matter of fact, if G is as in the Theorem, each of its countable subgroups satisfies (as sketched above) by [105] the Baum–Connes Conjecture and thus does G by 5.10.

The following is a simple example.

Corollary 5.14. *Let G be a group with $\operatorname{cd}_{\mathbb{Q}} G \leq 1$. Then G satisfies the* **BCC**.

Proof. A group G of rational cohomological dimension ≤ 1 admits a one-dimensional $\underline{E}G$, [40]. Since the stabilizers for the G-action on $\underline{E}G$ are all finite, they satisfy **BCC**. □

Here are some examples of groups with $\mathrm{cd}_{\mathbb{Q}} \leq 1$: finite groups, free groups, $\mathbb{Q}/\mathbb{Z}$ and finite extensions of free groups (the class of groups satisfying $\mathrm{cd}_{\mathbb{Q}} \leq 1$ is the same as the class of groups admitting a 1-dimensional $\underline{E}G$, see [40]).

There are two much larger classes of groups satisfying **BCC**, which we denote by $\mathbf{LH}\mathcal{TH}$ and $\mathbf{LH}\mathcal{ETH}$ respectively and which we define below. The **H** should indicate that these groups have a *hierarchical* structure, and the $\mathcal{TH}$ reminds that they are constructed out of groups acting on trees, with stabilizers having the *Haagerup property*. By closing $\mathbf{H}\mathcal{TH}$ with respect to certain *extensions* – indicated by $\mathcal{E}$ – we obtain $\mathbf{H}\mathcal{ETH}$. Finally, in view of 5.10 we can pass to even larger classes $\mathbf{LH}\mathcal{TH}$ and $\mathbf{LH}\mathcal{ETH}$ of groups G with the property that all finitely generated subgroups of G lie in $\mathbf{H}\mathcal{TH}$ resp. $\mathbf{H}\mathcal{ETH}$. The **L** indicates that the groups in question are *locally* in $\mathbf{H}\mathcal{TH}$ resp. $\mathbf{H}\mathcal{ETH}$.

Definition 5.15. (**Haagerup Property**) A group G is said to have the *Haagerup Property* (some authors say "G is a-T-menable"), if it admits a metrically proper isometric action on some (affine) Hilbert space.

The following is a simple observation.

Lemma 5.16. *Any discrete group with the Haagerup Property is countable.*

Proof. Recall that an isometric action of a group G on an affine Hilbert space $\mathcal{H}$ is called *metrically proper* if for all bounded subsets $B \subset \mathcal{H}$ the set $\{g \in G \,|\, gB \cap B \neq \emptyset\}$ is relatively compact in G which, in case of a discrete group means that these sets are finite. Thus, any orbit $Gx_0 \subset \mathcal{H}$ is countable, because it is contained in a countable union of balls $B(x_0, n)$ centered at x_0. Since the stabilizer G_{x_0} of x_0 is finite and $|G|=|G_{x_0}| \cdot |Gx_0|$, we conclude that a discrete group with the Haagerup Property is necessarily countable. □

For an extensive discussion of groups having the Haagerup property, the reader is referred to the book by Cherix, Cowling, Jolissaint, Julg and Valette [28]. Haagerup established this property for countable free groups (cf. [53]). The **BCC** has been proved for groups having the Haagerup property by Higson–Kasparov [59]. Examples of groups having the Haagerup property include: surface groups, countable amenable groups ([9]), Coxeter groups (cf. Bozejko, Januszkiewicz and Spatzier [15]), groups acting properly and isometrically on real or complex hyperbolic space, in particular, fundamental groups of not necessarily compact hyperbolic manifolds (cf. [47]), groups acting properly by isometries on a locally finite CAT(0) cubical complex (this is proved in Julg's survey [67], using a result of Niblo–Reeves [103]). It should be noted that the class of groups having the Haagerup property is certainly closed under passing to subgroups; it is also closed under passing to a finite extension: if G acts metrically properly by isometries on the Hilbert space $\mathcal{H}$ and G is a subgroup of finite index in some group L, then the coinduced L-action on the Hilbert space $\mathrm{Coind}_G^L \mathcal{H} = \mathrm{map}_G(L, \mathcal{H})$ is a metrically proper isometric action. Similarly, the product of finitely many groups having the Haagerup property has the Haagerup property. But a semi-direct product of groups having the Haagerup property need not have the Haagerup property: $\mathbb{Z}^2$ and $S\ell_2(\mathbb{Z})$ have the Haagerup property, but their semi-direct product $\mathbb{Z}^2 \rtimes S\ell_2(\mathbb{Z})$ does not (cf. [56]). More examples of groups which do not have the Haagerup property are given by the infinite groups having *Kazhdan's Property T* . Recall that a group has Kazhdan's property T, if every isometric action of G on an (affine) Hilbert space has a fixed point. This is known to be equivalent to the following: *For every unitary representation M of G one has* $H^1(G; M) = 0$; on the other hand, for infinite amenable groups G one has $H^1(G; \ell_2 G) \neq 0$: indeed, G is infinite and amenable if and only if $H^1(G; \ell_2 G)$ is not Hausdorff in its natural quotient topology coming from the Hilbert G-module of 1-cocycles cf. [52]. Thus, infinite amenable groups do not have Kazhdan's Property T. Obviously and more generally, if G has the Haagerup property, no infinite subgroup of G has property T. Examples of groups with the property T include $Sl(n, \mathbb{Z})$ for $n \geq 3$. There are very few groups with the property T for which the Baum–Connes Conjecture has been proved (e.g. cocompact lattices in $S\ell_3(\mathbb{R})$, or $Sp(n, 1)$, $n \geq 2$, see Lafforgue [74]); the latter groups are specially interesting since they have property T and are hyperbolic.

We can now define the class of groups $\mathbf{H}\mathcal{TH}$ in quite an analogous way as Kropholler's class $\mathbf{H}\mathfrak{F}$ (cf. [72]) is defined.

Definition 5.17. $\mathbf{H}\mathcal{TH}$ is the smallest class of groups containing all groups having the Haagerup property and containing a group G, if G admits a one-dimensional contractible G-CW-complex whose stabilizers are already in $\mathbf{H}\mathcal{TH}$. A group belongs to $\mathbf{LH}\mathcal{TH}$ if all of its finitely generated subgroups belong to $\mathbf{H}\mathcal{TH}$.

Another way of describing $\mathbf{H}\mathcal{TH}$ is as follows. Denote the class of groups having the Haagerup property by $\mathbf{H}_0\mathcal{TH}$. Then define by induction for each ordinal number α a class $\mathbf{H}_\alpha\mathcal{TH}$ as follows: for a successor ordinal $\beta + 1$, $\mathbf{H}_{\beta+1}\mathcal{TH}$ is the class of groups G which admit a one-dimensional contractible G-CW-complex whose stabilizers are in $\mathbf{H}_\beta\mathcal{TH}$; for a limit ordinal α one puts $\mathbf{H}_\alpha\mathcal{TH} = \cup_{\gamma<\alpha} \mathbf{H}_\gamma\mathcal{TH}$. The class $\mathbf{H}\mathcal{TH}$ is then given by

$$\mathbf{H}\mathcal{TH} = \bigcup_\alpha \mathbf{H}_\alpha\mathcal{TH}.$$

For instance, the semi-direct product $\mathbb{Z}^2 \rtimes S\ell_2(\mathbb{Z})$ lies in $\mathbf{H}_1\mathcal{TH}$: it acts, via its projection onto $S\ell_2(\mathbb{Z})$ on a tree with stabilizers which are finite extensions of $\mathbb{Z}^2$, which are amenable groups, thus lying in $\mathbf{H}_0\mathcal{TH}$; but, as mentioned earlier, $\mathbb{Z}^2 \rtimes S\ell_2(\mathbb{Z})$ does not have the Haagerup property and therefore does not already lie in $\mathbf{H}_0\mathcal{TH}$.

Clearly, the class $\mathbf{H}\mathcal{TH}$ is closed under amalgamation and HNN-extensions. It is also closed under passing to subgroups and, therefore, the class $\mathbf{LH}\mathcal{TH}$ contains the class $\mathbf{H}\mathcal{TH}$. Also, if G is the fundamental group of a graph of groups lying in $\mathbf{H}_\alpha\mathcal{TH}$, then G lies in $\mathbf{H}_{\alpha+1}\mathcal{TH}$. This implies that, for instance, *Haken* three-manifold groups (including knot groups) lie in $\mathbf{H}\mathcal{TH}$ (cf. [105]). Countable one-relator groups lie in $\mathbf{H}\mathcal{T}H$. This can be seen as follows. Such a group is a countable union of finitely generated one-relator groups. Let G be a finitely generated one-relator group, given by some presentation $\langle S|\, r\rangle$. If the relator r only involves ≤ 1 generator, the group G is a free product of a finitely generated free group and a cyclic group, thus has the Haagerup property (the free product of groups having the Haagerup Property has the Haagerup Property cf. [28]). If the relator involves $n > 1$ generators and has length $\ell \geq 2$, then as explained in Bieri's Queen Mary Notes [12], G is the fundamental group of a graph of subgroups of one-relator groups with relators of length less than ℓ. Thus, by induction $G \in \mathbf{H}\mathcal{TH}$. Since every finitely generated subgroup of an arbitrary one-relator group G is contained in a finitely generated one-relator subgroup of G, it follows that all one-relator groups lie in $\mathbf{LH}\mathcal{TH}$.

Theorem 5.18. *Let G be a group in $\mathbf{LH}\mathcal{TH}$. Then the Baum–Connes Conjecture holds for G and G is K_*-amenable. The class $\mathbf{LH}\mathcal{TH}$ has the following closure properties:*

- *it is closed under passing to subgroups*
- *it is closed under amalgamated products and HNN-extensions*
- *it is closed under arbitrary directed unions of subgroups.*

The class $\mathbf{LH}\mathcal{TH}$ contains in particular all

- *finite groups and free groups and, more generally, groups of rational cohomological dimension ≤ 1*
- *surface groups and, more generally, one-relator groups*
- *abelian groups and, more generally, amenable groups*
- *knot groups and, more generally, Haken three-manifold groups*
- *Coxeter groups, fundamental groups of hyperbolic manifolds and Thompson's group F.*

Proof. Using Oyono-Oyono's Theorem 5.13 and transfinite induction, we see that the groups in $\mathbf{LH}\mathcal{TH}$ all satisfy the Baum–Connes Conjecture. The groups in $\mathbf{H}_0\mathcal{TH}$ have the Haagerup property and, therefore, are K-amenable. By 5.6, if a group G admits a one-dimensional contractible G-CW-complex with K_*-amenable stabilizers, then G is K_*-amenable. Thus, again using transfinite induction, the groups in $\mathbf{LH}\mathcal{TH}$ are all K_*-amenable. The closure properties are obvious and the examples were discussed earlier, with the exception of the last one. Farley proved in [49] that Thompson's group F acts metrically properly by isometries on an affine Hilbert space and it therefore lies already in $\mathbf{H}_0\mathcal{TH}$. □

Recall that a group G is called *acyclic* if $H_*(G;\mathbb{Z}) = 0$ for $* > 0$. The class $\mathbf{LH}\mathcal{TH}$ is large enough to contain for every group G a surjection $\mathcal{T}(G) \to G$ with $\mathcal{T}(G) \in \mathbf{LH}\mathcal{TH}$ such that the kernel is acyclic, inducing therefore for every G-module M an isomorphism

$$H_*(\mathcal{T}(G); M) \to H_*(G; M).$$

This will be seen in the course of the proof of the following theorem, which can be paraphrased by saying that *torsion-free groups admit good approximations by K_*-amenable groups which satisfy the* **BCC**.

Theorem 5.19. *Let G be a not necessarily countable torsion-free group. Then there is a surjective group homomorphism $\phi : H \to G$ with H a torsion-free group satisfying the Baum–Connes Conjecture and being K_*-amenable (cf. Definition 5.5) so that there is a commutative diagram*

$$\begin{array}{ccccc} K_*^H(\underline{E}H) & \xrightarrow{\cong} & K_*^{top}(C^*(H)) & \xrightarrow[\alpha]{\cong} & K_*^{top}(C_r^*(H)) \\ \cong\downarrow \phi_* & & \downarrow \phi_*^{top} & & \downarrow \beta \\ K_*^G(\underline{E}G) & \longrightarrow & K_*^{top}(C^*(G)) & \xrightarrow{\gamma} & K_*^{top}(C_r^*(G)). \end{array}$$

The map β here is defined to be $\gamma \circ \phi_*^{top} \circ \alpha^{-1}$. Furthermore, if BG is finite-dimensional resp. finite, then H can be chosen in a way that BH is finite-dimensional resp. finite too; if G is countable, H can be chosen to be countable too.

Proof. Let G be an arbitrary group. We write $X = K(G,1)$ as a union of finite, connected subcomplexes $\{X_\alpha\}$, containing a chosen base-point $x_0 \in X$. Then by Kan–Thurston [68], one can find for each α a group G_α and a map $\phi : K(G_\alpha, 1) \to X_\alpha$ inducing a homology isomorphism with arbitrary $\pi_1(X_\alpha)$-module coefficients. This implies that $G_\alpha \to \pi_1(X_\alpha)$ is surjective. According to Baumslag–Dyer–Heller [7], one can even choose G_α to be geometrically finite, thus torsion-free and countable, with $K(G_\alpha, 1)$ a finite CW-complex (recall that a group is called geometrically finite, if it admits a classifying space which is a finite CW-complex). Checking their explicit construction [7], one sees that their G_α is constructed out of a finitely generated one-relator group, using amalgamations and HNN-extensions, thus staying in $\mathbf{H}\mathcal{TH}$: we may assume that G_α is torsion-free, countable, geometrically finite and lies in $\mathbf{H}\mathcal{TH}$. We now first complete the proof in case G is countable. We can then choose $X = K(G,1)$ to be a countable CW-complex and we can assume that the indexing set for the α's is $\mathbb{N}$, with $X_\alpha \subset X_{\alpha+1}$. The inductive construction of [7] shows that we may further assume that G_α is a subgroup of $G_{\alpha+1}$ so that $H := \operatorname{colim}_\alpha G_\alpha$ is torsion-free and it follows that $H \in \mathbf{H}\mathcal{TH}$, being a countable increasing union of groups in $\mathbf{H}\mathcal{TH}$. The construction yields a homomorphism $H := \operatorname{colim}_\alpha G_\alpha \to G$, inducing an isomorphism in homology with G-module coefficients, in particular with coefficients in $\mathbb{Z}$ and, therefore, in any homology theory defined by a spectrum, in particular in K-homology. Thus, in case of a countable torsion-free G

$$\begin{aligned} K_*^H(\underline{E}H) &\cong K_*(BH) \cong \operatorname{colim} K_*(K(G_\alpha,1)) \cong \operatorname{colim} K_*(X_\alpha) \cong K_*(BG) \\ &\cong K_*^G(\underline{E}G). \end{aligned}$$

Moreover, since $H = \operatorname{colim}_\alpha G_\alpha$ lies in $\mathbf{H}\mathcal{TH}$, H is K_*-amenable. In case BG is finite, $X = X_{\alpha_0}$ for a large $\alpha_0 \in \mathbb{N}$ so that $H = G_{\alpha_0}$ and BG_{α_0} is finite; similarly, in case BG is finite dimensional, BH is finite dimensional too. If the group G in question is not countable, the proof is very similar; the group H one ends up with then lies in **LH**$\mathcal{TH}$. □

Remark 5.20. By a result of Ian Leary [77] one can even choose the approximating groups G_α to be locally finite cubical CAT(0)-groups (meaning that they admit proper cocompact actions by isometries on locally finite cubical CAT(0) simplicial complexes).

In case of countable groups, there is an even stronger conjecture than **BCC**, which is known under the name *Baum–Connes Conjecture with Coefficients* and which we abbreviate with **BCPwC** (*Baum–Connes Property with Coefficients*): associated to a separable C^*-algebra A with an action of a countable G there is an assembly map, which we won't define in these notes, of the form

$$\Theta(G, A) : K_*^G(\underline{E}G; A) \to K_*^{top}(A \rtimes G).$$

Here $A \rtimes G$ stands for the reduced crossed-product C^*-algebra. The map $\Theta(G, A)$ reduces to our assembly map $\Theta(G)$ (as described 5.9) in case $A = \mathbb{C}$. The **BCPwC** for a group G is the statement that $\Theta(G, A)$ is an isomorphism for every A. A basic difference between **BCC** and **BCPwC** is that the latter is *subgroup closed*, cf. Chabert [24]: if **BCPwC** holds for G then it holds for all subgroups of G too; for more information concerning **BCPwC** the reader is referred to [105] and [106]. However, not every countable group has the *Baum–Connes Property with Coefficients*; for counterexamples, modulo a yet unpublished result due to Gromov, the reader is referred to the recent preprint [60] of Higson, Lafforgue and Skandalis. Because of the existence of these counterexamples, we talk about the Baum–Connes *Property with Coefficients* rather than the Baum–Connes *Conjecture with Coefficients*.

The countable groups in **H**$\mathcal{TH}$ actually all satisfy **BCPwC**. This follows from the inductive definition 5.17, together with the fact that the groups in $\mathbf{H}_0\mathcal{TH}$ satisfy **BCPwC** by [59], and that 5.13 holds for countable groups with the condition and conclusion "**BCC**" replaced by "**BCPwC**", see [105].

One can construct even more groups satisfying the Baum–Connes Conjecture using the following result due to Oyono-Oyono [106].

Theorem 5.21. *Let $\pi : G \to Q$ be a surjective group homomorphism with G countable. Assume that Q as well as the preimages $\pi^{-1}(F) < G$ for every finite subgroup $F < Q$ satisfy* **BCPwC**. *Then G satisfies* **BCPwC** *too.*

This leads us to consider the following class of groups, containing **LH**$\mathcal{TH}$.

Definition 5.22. Let **H**$\mathcal{ETH}$ be the smallest class of groups containing **H**$\mathcal{TH}$ and containing a group G if either G is countable and admits a surjective map $\pi : G \to Q$ with Q and $\pi^{-1}(F)$ in **H**$\mathcal{ETH}$ for every finite subgroup $F < Q$, or if G admits a one-dimensional contractible G-CW-complex with stabilizers all in **H**$\mathcal{ETH}$. The class **LH**$\mathcal{ETH}$ consists of all groups whose finitely generated subgroups lie in **H**$\mathcal{ETH}$.

Again, $\mathbf{H}\mathcal{ETH}$ could be defined as the union of classes $\mathbf{H}_\alpha\mathcal{ETH}$ as follows. Put $\mathbf{H}_0\mathcal{ETH}$=$\mathbf{H}\mathcal{TH}$ and define for ordinals α by induction new classes: $G \in \mathbf{H}_{\alpha+1}\mathcal{ETH}$ if either G is countable and admits a surjective homomorphism $\pi : G \to Q$ with Q and $\pi^{-1}(F)$ in $\mathbf{H}_\alpha\mathcal{ETH}$ for every finite subgroup $F < Q$, or G admits a one-dimensional contractible G-CW-complex with all stabilizers in $\mathbf{H}_\alpha\mathcal{ETH}$; for a limit ordinal β put $\mathbf{H}_\beta\mathcal{ETH} = \cup_{\alpha<\beta}\mathbf{H}_\alpha\mathcal{ETH}$. From this description and Theorem 5.21 it is clear that all groups in $\mathbf{H}\mathcal{ETH}$ satisfy the **BCC**, and all countable groups in $\mathbf{H}\mathcal{ETH}$ even satisfy **BCPwC**. It is clear from the definition that $\mathbf{H}\mathcal{ETH}$ is subgroup-closed and closed under passing to the fundamental group of a graph of groups.

Theorem 5.23. *All groups in* $\mathbf{LH}\mathcal{ETH}$ *satisfy the Baum–Connes Conjecture (cf.* 5.9*). The class of groups* $\mathbf{LH}\mathcal{ETH}$ *has the following properties:*

- *it contains the class* $\mathbf{LH}\mathcal{TH}$*, in particular all groups listed in Theorem* 5.18
- *it is closed under passing to subgroups*
- *it is closed under extensions with torsion-free quotients*
- *it closed under passing to the fundamental group of a graph of groups*
- *it is closed under finite products.*

Moreover, every countable group in $\mathbf{LH}\mathcal{ETH}$ *satisfies the Baum–Connes Property with Coefficients.*

Proof. Most of the claims are obvious in view of the definitions. For instance, to see that $G \times H$ belongs to $\mathbf{LH}\mathcal{ETH}$ if G and H do, one writes $G \times H$ as a directed union of finitely generated subgroups of the form $G_\alpha \times H_\alpha$ and notices that the projection $\pi : G_\alpha \times H_\alpha \to H_\alpha$ has the property that $\pi^{-1}(F)$ lies in $\mathbf{H}\mathcal{ETH}$ for $F \subset H_\alpha$, and the result follows. We used here the fact that if a countable group C lies in $\mathbf{H}\mathcal{ETH}$ and F is a finite group, then $C \times F$ lies in $\mathbf{H}\mathcal{ETH}$ too as one see from the exact sequence $F \to C \times F \to C$ by observing that a counterimage of a finite subgroup of C in $C \times F$ is finite and has therefore the **BCPwC**. The reader can check the other claims easily. □

Theorem 5.24. *Let* G *be a countable group. There are natural embeddings of* G *in groups* $\alpha(G)$*,* $\beta(G)$ *and* $\gamma(G)$ *such that in* $\alpha(G)$ *and* $\gamma(G)$ *all elements of infinite order are conjugate, and the groups* $\beta(G)$*,* $\gamma(G)$ *are acyclic. Moreover, the constructions satisfy the following.*

- *The groups* $\alpha(G)$*,* $\beta(G)$ *and* $\gamma(G)$ *are countable; they are torsion-free if* G *is torsion-free.*
- *If* G *satisfies* **BCC** *then so does* $\alpha(G)$*.*
- *If* G *is* K*-amenable then so are* $\alpha(G)$*,* $\beta(G)$ *and* $\gamma(G)$*.*
- *If* $G \in \mathbf{LH}\mathcal{ETH}$ *then so are* $\alpha(G)$*,* $\beta(G)$ *and* $\gamma(G)$*.*

Proof. First, we describe the construction of $\alpha(G)$. If $x, y \in G$ are two elements of infinite order, we can form the HNN-extension

$$G(x,y) := \langle G, t \,|\, txt^{-1} = y\rangle$$

containing G as a subgroup. Repeating this construction for all pairs of elements of infinite order coming from G, we obtain an inclusion $G \to \alpha_1(G)$ such that in $\alpha_1(G)$ all elements of infinite order coming from G are conjugate. Putting $\alpha_{n+1}(G)$ equal to $\alpha_1(\alpha_n(G))$ and defining

$$\alpha(G) := \bigcup_{n\geq 1} \alpha_n(G)$$

we obtain a group with the desired properties. For $\beta(G)$ we use another standard construction, providing an embedding of G in a binate group (for a discussion of this construction as well as a definition of the term *binate*, see J. Berrick's paper [10]). Consider the following HNN-extension of $G \times G$, containing G via the left embedding $G \to G \times G, g \mapsto (g, 1)$:

$$\beta_1(G) := \langle G \times G, t \,|\, t(1,g)t^{-1} = (g,g), \forall g \in G\rangle.$$

Putting $\beta_{n+1}(G) = \beta_1(\beta_n(G))$ and considering $\beta_n(G)$ as a subgroup of $\beta_{n+1}(G)$ via $x \mapsto (x, 1)$, one defines

$$\beta(G) := \bigcup_{n\geq 1} \beta_n(G)$$

which is a countable acyclic group. The construction of $\gamma(G)$ uses a combination of the previous two constructions as follows; we define $\gamma_1(G)$ to be $\beta(\alpha(G))$, and inductively $\gamma_{n+1}(G) = \gamma_1(\gamma_n(G))$ to obtain

$$\gamma(G) := \bigcup_{n\geq 1} \gamma_n(G),$$

an acyclic group with all elements of infinite order conjugate. It follows from well-known properties of HNN-extensions that for G torsion-free, the groups $\alpha(G)$, $\beta(G)$ and $\gamma(G)$ are torsion-free too. If G satisfies **BCC** then so does $\alpha(G)$, since **BCC** is closed under HNN-extensions and directed unions. For the statement concerning K-amenability we use that the cartesian product of two K-amenable groups is K-amenable [30], and that the fundamental group of a countable graph of countable K-amenable groups is K-amenable too [109]. To show that for G in $\mathbf{LH}\mathcal{ETH}$ the groups $\alpha(G)$, $\beta(G)$ and $\gamma(G)$ are in $\mathbf{LH}\mathcal{ETH}$, we use Theorem 5.23. □

To conclude the discussion of families of groups satisfying **BCC**, we want to show, following an idea of Schick, that the Artin braid groups B_n lie in $\mathbf{H}\mathcal{ETH}$; we don't know whether (for $n \geq 4$) B_n lies already in $\mathbf{H}\mathcal{TH}$.

Theorem 5.25. **(Schick** [118]**)** *The Artin braid groups B_n lie in $\mathbf{H}\mathcal{ETH}$ and satisfy therefore the Baum-Connes Conjecture.*

Proof. Consider the extension $P_n \rightarrowtail B_n \twoheadrightarrow \Sigma_n$ with P_n denoting the pure braid group on n strings. The *Artin combing* presents the pure braid group as an iterated semi-direct product

$$P_n = F_{n-1} \rtimes P_{n-1} = F_{n-1} \rtimes F_{n-2} \rtimes \cdots \rtimes F_1$$

with F_i free of rank i. It is clear by induction, that $P_n \in \mathbf{H}\mathcal{ETH}$. We want to construct for a fixed n a new extension $H \rightarrowtail B_n \twoheadrightarrow Q$ with Q torsion-free in $\mathbf{H}\mathcal{ETH}$ and $H < P_n$. It then follows that B_n lies in $\mathbf{H}\mathcal{ETH}$ and, therefore, satisfies the Baum–Connes Conjecture. For the construction of H we look at the i-th quotient $P_n/\Gamma^i P_n$ of the lower central series of P_n which, by a result of Falk and Randell [46], agrees with $\prod_{k=1}^{n-1} F_k/\Gamma^i(\prod_{k=1}^{n-1} F_k)$ so that by well-known results on the lower central series quotients for free groups, each quotient $H_i := P_n/\Gamma^i P_n$ is a finitely generated torsion-free nilpotent group, fitting into a commutative diagram

$$\begin{array}{ccccc} P_n & \longrightarrow & B_n & \xrightarrow{\pi} & \Sigma_n \\ \downarrow & & \downarrow & & \| \\ H_i & \longrightarrow & K_i & \longrightarrow & \Sigma_n, \end{array}$$

with $K_i := B_n/\Gamma^i P_n$. Note also that $\bigcap_i \Gamma^i P_n = \{e\}$ by [46]. Moreover, the result on the lower central series of P_n implies that the short exact sequence

$$F_{n-1} \rightarrowtail P_n \twoheadrightarrow P_{n-1}$$

yields a (split) injection $H_1(F_{n-1};\mathbb{Z}) \to H_1(P_n;\mathbb{Z})$ and therefore the action of P_{n-1} on $H_*(F_{n-1};\mathbb{Z})$ is trivial; later on, we will use that, as a consequence, P_{n-1} acts nilpotently (even trivially) on $H^*(F_{n-1};\mathbb{Z}/p\mathbb{Z})$. We claim that $K_i = B_n/\Gamma^i P_n$ is torsion-free for i large enough. As H_i is torsion-free, K_i has certainly no p-torsion for primes $p > n$. A standard argument, using induction on the nilpotency length of H_i, shows that for a fixed prime p the number of conjugacy classes of elements of order p in K_i is finite. We now claim that K_i is p-torsion-free for $i \gg 0$. Suppose not: then there is a compatible tower, with $B_n(p) := \pi^{-1}(\mathbb{Z}/p\mathbb{Z}) < B_n$, where $\mathbb{Z}/p\mathbb{Z} \subset \Sigma_n$:

$$\begin{array}{ccccc} P_n & \longrightarrow & B_n(p) & \longrightarrow & \mathbb{Z}/p\mathbb{Z} \\ \vdots & & \vdots & & \vdots \\ \downarrow & & \downarrow & & \| \\ H_{i+1} & \longrightarrow & K_{i+1}(p) & \xrightarrow{\text{split}} & \mathbb{Z}/p\mathbb{Z}, \\ \downarrow & & \downarrow & & \| \\ H_i & \longrightarrow & K_i(p) & \xrightarrow{\text{split}} & \mathbb{Z}/p\mathbb{Z}, \end{array}$$

and therefore, by the argument indicated earlier each $K_i(p)$ contains finitely many conjugacy classes of elements of order p only, meaning that there is an element of order p in $\lim_i K_i(p)$. Recall that a group G is called p-good, if the map $G \to G_p^\wedge$ to the p-profinite completion of G induces an isomorphism

$$H^*_{\text{cont}}(G_p^\wedge;\mathbb{Z}/p\mathbb{Z}) \to H^*(G;\mathbb{Z}/p\mathbb{Z}),$$

where H^*_{cont} stands for *continuous cohomology* (cf. Serre [123]). A Serre spectral sequence argument shows that if G fits into a short exact sequence

$$H \rightarrowtail G \twoheadrightarrow Q$$

with H and Q p-good, $H^*(H;\mathbb{Z}/p\mathbb{Z})$ finite dimensional and Q acting nilpotently on $H^*(H;\mathbb{Z}/p\mathbb{Z})$ then G is p-good too. This implies in particular that the pure braid groups P_n are p-good, being iterated extensions of finitely generated free groups and, as observed, with $P_n/F_{n-1} = P_{n-1}$ acting trivially on $H^*(F_{n-1};\mathbb{Z}/p\mathbb{Z})$. Also, since $H^*(P_n;\mathbb{Z}/p\mathbb{Z})$ is finite dimensional the finite extension $B_n(p)$ of P_n is p-good too, since the p-group $B_n(p)/P_n$ can only act in a nilpotent way on $H^*(P_n;\mathbb{Z}/p\mathbb{Z})$. Since all finite p-groups are nilpotent, one has

$$(P_n)^\wedge_p = (\lim_i H_i)^\wedge_p = \lim_i((H_i)^\wedge_p)$$

and similarly

$$B_n(p)^\wedge_p = \lim_i(K_i(p)^\wedge_p)$$

so that, $B_n(p)$ being a subgroup of B_n, it has finite cohomological dimension,

$$H^*(B_n(p);\mathbb{Z}/p\mathbb{Z}) \cong H^*_{\text{cont}}(B_n(p)^\wedge_p;\mathbb{Z}/p\mathbb{Z}) = 0, \text{ for } * \gg 0.$$

But this implies that the group

$$B_n(p)^\wedge_p = \lim_i\,(K_i(p)^\wedge_p)$$

is p-torsion-free, which contradicts our assumption (we used here the fact that for a pro-p-group G with $\operatorname{vcd}_p G < \infty$ and $H^*_{\text{cont}}(G;\mathbb{Z}/p\mathbb{Z}) = 0$ for $* \gg 0$, the group G must be p-torsion-free; indeed, if $C < G$ is a subgroup of order p, then the restriction map $H^{2i}_{cont}(G;\mathbb{Z}/p\mathbb{Z}) \to H^{2i}(C;\mathbb{Z}/p\mathbb{Z})$ is non-zero for some $i > 0$, which, using the ring structure, shows that for any n there is a $j > n$ with $H^{2j}_{cont}(G;\mathbb{Z}/pZ)$ non-zero). Therefore, for $i \gg 0$ one has an exact sequence

$$\Gamma^i P_n \to B_n \to K_i$$

with K_i a torsion-free finite extension of a nilpotent group, so that K_i is amenable and lies therefore in $\mathbf{H}_0\mathcal{TH}$. On the other hand $\Gamma_i P_n \in \mathbf{H}\mathcal{TH}$, being a subgroup of P_n, which completes our proof. □

Remark 5.26. There are some interesting general results concerning *extensions* of groups satisfying **BCC**, see Chabert [22] and [23] as well as Oyono-Oyono [106].

A useful device for computing $K^G_*(\underline{E}G)$ is the Atiyah–Hirzebruch spectral sequence. By considering $K^G_i(\,?\,)$ as an object in G-$\text{Mod}_{\mathfrak{Fin}}$ one has for a proper G-CW-complex X the associated Bredon homology groups $H^{\mathfrak{Fin}}_p(X;K^G_q(\,?\,))$, which are obtained as homology groups of the chain complex $\underline{C_*(X)} \otimes_{\mathfrak{Fin}} K^G_*(\,?\,)$. Using the skeleton filtration $\{X^p\}_{p\in\mathbb{N}}$ of X one obtains a spectral sequence with

$$E^1_{p,q} = K^G_{p+q}(X^p, X^{p-1})$$

and differential

$$d^1_{p,q} : E^1_{p,q} \longrightarrow E^1_{p-1,q}$$

induced by the long exact sequence of the triple (X^{p+1}, X^p, X^{p-1}) in K^G_*-homology. The resulting spectral sequence

$$E^2_{p,q} = H^{\mathfrak{Fin}}_p(X; K^G_q(\,?\,)) \Rightarrow K^G_{p+q}(X)$$

lives in the first and forth quadrant and its differential d^n on $E^n_{*,*}$ has bidegree $(-n, n-1)$. In case X is finite dimensional, this implies that $E^n_{p,q} = E^\infty_{p,q}$ for $n > \dim X$. The E^∞ term is then the associated graded group of a finite filtration $\{F_{p,q}K^G_{p+q}(X); 0 \le p \le \dim X\}$ with

$$F_{p-1,q+1}K^G_{p+q}(X) \subset F_{p,q}K^G_{p+q}(X); \quad \text{and}$$

$$F_{p,q}K^G_{p+q}(X)/F_{p-1,q+1}K^G_{p+q}(X) \cong E^\infty_{p,q}.$$

By definition, $F_{p,q}K^G_{p+q}(X)$ is the image of the natural map $K^G_{p+q}(X^p) \to K^G_{p+q}(X)$. Note also that for n even, the differential d^n is zero, since either its source or its target vanishes (for a finite group H and q odd, $K^G_q(G/H) \cong K^{top}_q(\mathbb{C}[H])$ vanishes). Looking at

$$F_{0,0}K^G_0(X) \subset F_{1,-1}K^G_0(X) \subset \cdots \subset K^G_0(X)$$

one always has $F_{0,0}K^G_0(X) = E^\infty_{0,0}$, which is a factor group of $E^2_{0,0}$. Similarly

$$F_{0,1}K^G_1(X) \subset F_{1,0}K^G_1(X) \subset \cdots \subset K^G_1(X),$$

but here $F_{0,1}K^G_1(X) = 0$, since it is a quotient of the trivial group $H^{\mathfrak{Fin}}_0(X; K^G_1(\,?\,))$. As a result $F_{1,0}K^G_1(X) = E^\infty_{1,0}$, which is a factor group of $E^2_{1,0}$. The *edge homomorphisms* are the resulting natural maps

$$E^2_{0.0} = H^{\mathfrak{Fin}}_0(X; R_{\mathbb{C}}) \twoheadrightarrow E^\infty_{0,0} \subset K^G_0(X)$$

and

$$E^2_{1.0} = H^{\mathfrak{Fin}}_1(X; R_{\mathbb{C}}) \twoheadrightarrow E^\infty_{1,0} \subset K^G_1(X)$$

which correspond in case of $X = \underline{E}G$ to natural maps

$$\Lambda_0(G) : H^{\mathfrak{Fin}}_0(G; R_{\mathbb{C}}) = R^{\mathfrak{Fin}}_{\mathbb{C}}(G) \longrightarrow K^G_0(\underline{E}G)$$

and

$$\Lambda_1(G) : H^{\mathfrak{Fin}}_1(G; R_{\mathbb{C}}) \longrightarrow K^G_1(\underline{E}G).$$

Note that if $\dim \underline{E}G \le 2$, $\Lambda_0(G)$ is injective, because then $E^2_{0,0} = E^\infty_{0,0}$; we use here the fact that the differential d^r on $E^r_{*,*}$ has bidegree $(-r, r-1)$. If $\dim \underline{E}G \le 4$, the map $\Lambda_1(G)$ is injective, as then $E^2_{1,0} = E^\infty_{1,0}$. By looking at the other edge, one finds if $\dim X < \infty$, choosing $2n \ge \dim X$:

$$F_{2n,0}K^G_0(X) = K^G_0(X) \twoheadrightarrow E^\infty_{2n,0} \subset E^2_{2n,0} = H^{\mathfrak{Fin}}_{2n}(X; R_{\mathbb{C}})$$

and, since $F_{2n,1-2n}K^G_1(X) = F_{2n-1,2-2n}K^G_1(X) = K^G_1(X)$,

$$K^G_1(X) = F_{2n-1,2-2n}K^G_1(X) \twoheadrightarrow E^\infty_{2n-1,2-2n} \subset E^2_{2n-1,2-2n} = H^{\mathfrak{Fin}}_{2n-1}(X; R_{\mathbb{C}})$$

which for $X = \underline{E}G$ of dimension $\le 2n$ yields natural maps

$$\Gamma_{2n}(G) : K^G_0(\underline{E}G) \longrightarrow H^{\mathfrak{Fin}}_{2n}(G; R_{\mathbb{C}})$$

and

$$\Gamma_{2n-1}(G) : K_1^G(\underline{E}G) \longrightarrow H_{2n-1}^{\mathfrak{F}\mathrm{in}}(G; R_{\mathbb{C}}).$$

In particular, for proper G-CW-complexes of dimension 2 one can capture their entire equivariant K-homology in terms of Bredon homology groups, because then $E^2 = E^\infty$.

Theorem 5.27. *Let G be an arbitrary group such that $\dim \underline{E}G \leq 2$. Then there is a natural short exact sequence*

$$0 \longrightarrow R_{\mathbb{C}}^{\mathfrak{F}\mathrm{in}}(G) \xrightarrow{\Lambda_0(G)} K_0^G(\underline{E}G) \xrightarrow{\Gamma_2(G)} H_2^{\mathfrak{F}\mathrm{in}}(G; R_{\mathbb{C}}) \longrightarrow 0$$

and a natural isomorphism induced by the inverse pair of isomorphisms $\Lambda_1(G)$ and $\Gamma_1(G)$:

$$H_1^{\mathfrak{F}\mathrm{in}}(G; R_{\mathbb{C}}) \cong K_1^G(\underline{E}G).$$

Proof. Since $\dim \underline{E}G \leq 2$ one finds

$$F_{0,0} = F_{1,-1} \subset F_{2,-2} = K_0^G(\underline{E}G)$$

with $F_{0,0} = R_{\mathbb{C}}^{\mathfrak{F}\mathrm{in}}(G)$, and with $F_{2,-2}$ mapping onto $E_{2,-2}^\infty \cong H_2^{\mathfrak{F}\mathrm{in}}(G; R\mathbb{C})$, the kernel being $F_{1,-1}$. For $K_1^G(\underline{E}G)$ one finds

$$F_{1,0} = F_{2,-1} = K_1^G(\underline{E}G)$$

so that $H_1^{\mathfrak{F}\mathrm{in}}(G; R_{\mathbb{C}}) \cong F_{1,0} = K_1^G(\underline{E}G)$. □

Taking for G a one-relator group, we have $\dim \underline{E}G \leq 2$ and we can apply Lemma 3.23. In accordance with with Béguin–Bettaieb–Valette [8] we then find the following.

Corollary 5.28. *Let G be one-relator group with maximal finite subgroup $C < G$. Then there is a split short exact sequence*

$$R_{\mathbb{C}}(C) \rightarrowtail K_0^G(\underline{E}G) \twoheadrightarrow H_2(G; \mathbb{Z})$$

and natural isomorphisms

$$G_{ab}/\overline{C} \cong H_1^{\mathfrak{F}\mathrm{in}}(G; R_{\mathbb{C}}) \cong K_1^G(\underline{E}G),$$

where $\overline{C}$ denotes the image of C in G_{ab}.

Examples The following are two examples of torsion-free one-relator groups. Firstly, let $G = \langle x_1, \cdots, x_g | x_1^2 x_2^2 \cdots x_g^2 \rangle$, the fundamental group of a non-orientable surface of genus $g > 1$. Then

$$K_0^G(\underline{E}G) \cong \mathbb{Z}, \quad K_1^G(\underline{E}G) \cong \mathbb{Z}^{g-1} \oplus (\mathbb{Z}/2\mathbb{Z}).$$

Secondly, let $H = \langle x, y | xyx^{-1}y^{-2} \rangle$, a Baumslag–Solitar group. Then

$$K_0^G(\underline{E}G) \cong \mathbb{Z}, \quad K_1^G(\underline{E}G) \cong \mathbb{Z}.$$

Returning to the general case, we observe that the elements of $E^2_{0,q}$ are permanent cycles. The associated edge homomorphism takes the form

$$\Lambda_0(G) : H_0^{\mathfrak{Fin}}(\underline{E}G; R_{\mathbb{C}}) = R_{\mathbb{C}}^{\mathfrak{Fin}}(G) \longrightarrow K_0^G(\underline{E}G).$$

The kernel of $\Lambda_0(G)$ consists of the subgroup of $E^2_{0,*}$ generated by all images of higher differentials; it will follow from our discussion of the equivariant Chern character in the next section that the kernel of $\Lambda_0(G)$ is a torsion group. Because of dimension reasons, this kernel, being the image of higher differentials, is zero if $\dim \underline{E}G \leq 2$, as we already observed. If $\dim \underline{E}G = 3$, one has an exact sequence

$$H_3^{\mathfrak{Fin}} \xrightarrow{d^3} H_0^{\mathfrak{Fin}} \longrightarrow K_0^G \longrightarrow H_2^{\mathfrak{Fin}} \longrightarrow 0$$

by looking at $\{F_{0,2} \subset F_{2,0} = K_0^G\}$, and

$$0 \longrightarrow H_1^{\mathfrak{Fin}} \longrightarrow K_1^G \longrightarrow H_3^{\mathfrak{Fin}} \xrightarrow{d^3} H_0^{\mathfrak{Fin}}$$

coming from $\{F_{1,2} \subset F_{3,0} = K_1^G\}$. Hence, the previous theorem admits the following generalization.

Theorem 5.29. *Let G be an arbitrary group and assume that there exists a three-dimensional model for $\underline{E}G$. Then there is a natural 6-term exact sequence*

$$\begin{array}{ccccccc} 0 & \longrightarrow & H_1^{\mathfrak{Fin}}(G; R_{\mathbb{C}}) & \xrightarrow{\Lambda_1(G)} & K_1^G(\underline{E}G) & \xrightarrow{\Gamma_3(G)} & H_3^{\mathfrak{Fin}}(G; R_{\mathbb{C}}) \\ & & & & & & \big\downarrow d^3 \\ 0 & \longleftarrow & H_2^{\mathfrak{Fin}}(G; R_{\mathbb{C}}) & \xleftarrow{\Gamma_2(G)} & K_0^G(\underline{E}G) & \xleftarrow{\Lambda_0(G)} & H_0^{\mathfrak{Fin}}(G; R_{\mathbb{C}}) \end{array}$$

For related results and a thorough discussion of low-dimensional group homology in connection with the Baum–Connes assembly map, the reader is referred to Matthey's Thesis [95].

Note that one could describe $\Lambda_0(G)$ also without referring to the spectral sequence, as being the map induced from the G-homotopy classes $G/H \to \underline{E}G$, by taking the colimit of the associated maps $K_0^G(G/H) = R_{\mathbb{C}}(H) \to K_0^G(\underline{E}G)$ in equivariant K-theory.

6. The Equivariant Chern Character

The Chern character considered earlier in the non-equivariant setting admits the following generalization (see Lück [88]).

Theorem 6.1. *Let (X, A) be a pair of proper G-CW-complexes. Then there is a decomposition*

$$Ch_*^G : \bigoplus_{i \in \mathbb{Z}} H_{*+2i}^{\mathfrak{Fin}}(X, A; \mathbb{Q} \otimes R_{\mathbb{C}}) \xrightarrow{\cong} K_*^G(X, A) \otimes \mathbb{Q}.$$

For the groups on the left hand side one has isomorphisms

$$H_k^{\mathfrak{Fin}}(X, A; \mathbb{Q} \otimes R_{\mathbb{C}}) \cong \bigoplus_{[g] \in \mathrm{FC}(G)} H_k((X^g, A^g)/C_G(g); \mathbb{Q}), \quad k \in \mathbb{N},$$

the sum being taken over the set FC(G) *of conjugacy classes of elements of finite order in* G.

Proof. We will restrict ourselves to the case of an empty subspace A to simplify the notation. As we have already seen, there is a splitting of the covariant functor $\mathbb{Q} \otimes R_{\mathbb{C}}$ as a sum of functors

$$G/H \mapsto e_{S,H}(\mathbb{Q} \otimes R_{\mathbb{C}}(H)) = R_{\sharp,S}(G/H)$$

yielding (cf. Remark 3.26)

$$H_k^{\mathfrak{Fin}}(X;\mathbb{Q} \otimes R_{\mathbb{C}}) = \bigoplus_{[S] \in Z(G)} H_k(X^S/C_G(S);\mathbb{Q}) \otimes_{W_G(S)} R_{\sharp,S}(G/S)$$

with S running over the set $Z(G)$ of conjugacy classes of finite cyclic subgroups of G. The terms on the right-hand side decompose into a sum of terms of the form $H_k(X^g/C_G(g);\mathbb{Q})$, indexed by the set of G-conjugacy classes of generators g of the cyclic group S (cf. Theorem 3.27). To define Ch_n^G it suffices therefore to define for every finite cyclic subgroup $S < G$ and every $p \geq 0$ natural maps

$$\chi_p(S) : H_p(X^S/C_G(S);\mathbb{Q}) \otimes R_{\mathbb{C}}(S) \to K_p^G(X) \otimes \mathbb{Q}$$

satisfying the obvious compatibility conditions, which we don't check here, including $\chi_p(S)(wx \otimes y) = \chi_p(x \otimes wy)$ for $w \in W_G(S)$. The map $\chi_p(S)$ is defined as a composite of six homomorphisms α_i, $1 \leq i \leq 6$. We now describe these homomorphisms. We view $EG \times X^S$ as a $C_G(S)$-space by the diagonal action $g(e,x) = (ge, gx)$ so that the projection

$$\pi : EG \times_{C_G(S)} X^S \to X^S/C_G(S)$$

is an $H_*(\,?\,;\mathbb{Q})$ isomorphism. Indeed, for $\sigma \in X^S/C_G(S)$, $\pi^{-1}(\sigma)$ has the form $EG \times_{C_G(S)} C_G(S) \cdot \tilde{\sigma}$, with $\tilde{\sigma} \in X^S$ a lift of σ. Thus $C_G(S) \cdot \tilde{\sigma} \cong C_G(S)/H$ with H the stabilizer of $\tilde{\sigma}$, and therefore

$$\pi^{-1}(\sigma) \cong EG \times_{C_G(S)} (C_G(S)/H) \cong EG/H \simeq K(H,1),$$

which is a $\mathbb{Q}$-acyclic space. This implies that π is an $H_*(\;;\mathbb{Q})$-isomorphism. The map α_1 is now defined as the inverse of the natural isomorphism induced by π,

$$\pi_* : H_p(EG \times_{C_G(S)} X^S;\mathbb{Q}) \otimes R_{\mathbb{C}}(S) \to H_p(X^S/C_G(S);\mathbb{Q}) \otimes R_{\mathbb{C}}(S).$$

The map α_2 is the inverse of the ordinary Hurewicz isomorphism

$$\pi_p^{st}((EG \times_{C_G(S)} X^S)_+) \otimes \mathbb{Q} \otimes R_{\mathbb{C}}(S) \to H_p(EG \times_{C_G(S)} X^S;\mathbb{Q}) \otimes R_{\mathbb{C}}(S)$$

discussed in Section 4. The map

$$\alpha_3 : \pi_p^{st}((EG \times_{C_G(S)} X^S)_+) \otimes \mathbb{Q} \otimes R_{\mathbb{C}}(S) \to K_p^S(EG \times_{C_G(S)} X^S) \otimes \mathbb{Q}$$

is given by $\alpha_3(u \otimes 1 \otimes v) = u_*(v)$ where for $u \in \pi_p^{st}$, represented by $u : S^{i+p} \to S^i \wedge (EG \times_{C_G(S)} X^S)_+$ considered as an S-map with respect to trivial S-action, the map u_* is the induced map in the diagram

$$\begin{array}{ccc} R_{\mathbb{C}}(S) \otimes \mathbb{Q} = K_0^S(S^0,*) \otimes \mathbb{Q} & \xrightarrow{\cong} & K_{i+p}^S(S^{i+p},*) \otimes \mathbb{Q} \\ {\scriptstyle u_*}\downarrow & & \downarrow {\scriptstyle K_{i+p}^S(u)} \\ K_p^S(EG \times_{C_G(S)} X^S) \otimes \mathbb{Q} & \xleftarrow{\cong} & K_{i+p}^S(S^i \wedge (EG \times_{C_G(S)} X^S)_+,*) \otimes \mathbb{Q}. \end{array}$$

Now we view $EG \times X^S$ as a $(C_G(S) \times S)$-space with diagonal $C_G(S)$-action as before, and trivial action by the second factor S. The map, α_4 is now the inverse of the natural isomorphism

$$K_p^{C_G(S) \times S}(EG \times X^S) \otimes \mathbb{Q} \to K_p^S(EG \times_{C_G(S)} X^S) \otimes \mathbb{Q}$$

obtained by dividing out the free $C_G(S)$-action on $EG \times X^S$. By inducing the $(C_G(S) \times S)$-action on $EG \times X^S$ up to a G-action using the map $m_S : C_G(S) \times S \to G$, $(g,s) \mapsto gs$, one obtains the isomorphism

$$\alpha_5 : K_p^{C_G(S) \times S}(EG \times X^S) \otimes \mathbb{Q} \to K_p^G(\mathrm{Ind}_{m_S}(EG \times X^S)) \otimes \mathbb{Q}.$$

We used here Corollary 5.8 and the fact that the kernel of m_S, which is isomorphic to a diagonal copy of S, acts freely on $EG \times X^S$. The projection $\mathrm{Ind}_{m_S}(EG \times X^S) \to \mathrm{Ind}_{m_S} X^S$ composed with the natural G-map $G \times_{m_S} X^S \to X$ induces the last map α_6, mapping to $K_p^G(X) \otimes \mathbb{Q}$. The map Ch_*^G extends to a morphism of G-homology theories on proper G-CW-complexes and induces on orbits G/H, $H < G$, an isomorphism $R_{\mathbb{C}}(H) \otimes \mathbb{Q} \to R_{\mathbb{C}}(H) \otimes \mathbb{Q}$, thus inducing an isomorphism on all proper G-CW-complexes. □

Corollary 6.2. *For an arbitrary group G, there are natural injective maps*

$$H_{even}(BG;\mathbb{Q}) = \bigoplus_{i \in \mathbb{Z}} H_{2i}(BG;\mathbb{Q}) \to K_0^G(\underline{E}G) \otimes \mathbb{Q}$$

and

$$H_{odd}(BG;\mathbb{Q}) = \bigoplus_{i \in \mathbb{Z}} H_{2i+1}(BG;\mathbb{Q}) \to K_1^G(\underline{E}G) \otimes \mathbb{Q}.$$

Proof. As we have seen in the course of the proof of Theorem 6.1, for every proper G-CW-complex X and finite cyclic subgroup $S < G$, there is a natural injective map

$$\bigoplus_i H_{*+2i}(X^S/C_G(S);\mathbb{Q}) \otimes_{W_G(S)} R_{\sharp,S}(G/S) \to K_*^G(X) \otimes \mathbb{Q},$$

which is induced by the maps $\chi_{*+2i}(S)$ defined earlier. In case $X = EG$ and $S = \{e\}$ one has $R_{\sharp,S}(G/S) = \mathbb{Q}$ and the resulting map defines the map in question. □

7. Appendix: Related Conjectures

We will sketch in this section a few conjectures in algebra and topology, which are related to the Baum–Connes Conjecture 5.9. The proofs are only sketched and references are given to more comprehensive sources. For the convenience of the reader, we will introduce some short notations for the various conjectures. If it is known that a conjecture, when it holds for a certain group, also holds for all its subgroups, we say that the conjecture is *subgroup closed.*

BCC: Baum–Connes Conjecture

- *The assembly map $K_*^G(\underline{E}G) \to K_*^{top}(C_r^*(G))$ is an isomorphism.*

It is not known whether **BCC** is subgroup closed. As mentioned earlier, in the literature one finds a related more general statement, which is often referred to as the *Baum–Connes Conjecture with Coefficients*; for information on this conjecture, which – unfortunately – some authors just call the "Baum–Connes Conjecture" and which we won't discuss here, the reader is referred to [130]. Since this more general statement admits counterexamples, we call it **BCPwC**, the *Baum–Connes Property with Coefficients* (see the remarks on page 45). It is proved in [24] that **BCPwC** is subgroup closed.

If $x \in G$ has finite order $n > 1$, the element $\sum_{i=1}^{n} x^i/n$ defines a non-trivial idempotent in $\mathbb{Q}[G]$. A classical conjecture due to Kaplansky states that for a *torsion-free* group G the group algebra $\mathbb{Q}[G]$ has no idempotent besides of 0 and 1.

IC: Idempotent Conjecture

- *For a torsion-free group G, $\mathbb{Q}[G]$ has no non-trivial idempotent.*

More generally, Kadison conjectured that for a torsion-free group G, the reduced C^*-algebra $C_r^*(G)$ has no non-trivial idempotent either; we call this the *Strong Idempotent Conjecture* and denote it by

SIC: Strong Idempotent Conjecture

- *For a torsion-free group G, $C_r^*(G)$ has no non-trivial idempotent.*

It is easy to check that the strong idempotent conjecture holds for a finitely generated free abelian group. Indeed, $C_r^*(\mathbb{Z}^n)$ is isomorphic to the C^*-algebra of continuous $\mathbb{C}$-valued functions $C(T^n)$ on the torus $T^n = (S^1)^n$ and continuous idempotent functions $T^n \to \mathbb{C}$ only take the values 0 and 1 so that by continuity the 0-function and the 1-function are the only two idempotents.

The strong idempotent conjecture follows for a given group G, provided the *Kaplansky Trace*

$$\kappa_{C_r^*} : K_0^{top}(C_r^*(G)) \to \mathbb{R}$$

is integer valued. For the convenience of the reader, we recall the definition of $\kappa_{C_r^*}$. Write $\ell_2(G)$ for the complex Hilbert space of square summable functions $\phi : G \to \mathbb{C}$, with the group elements as a Hilbert basis ($g \in G$ is considered as a function $g : G \to \mathbb{C}$ by $g(x) = \delta_{g,x}$ so that $G \subset \ell_2(G)$; we can then write ϕ in the form $\sum\langle\phi, g\rangle g)$. There is a (continuous) embedding

$$C_r^*(G) \to \ell_2(G)$$

given by $f \mapsto f(e)$ with $e \in G$ the neutral element, considered as $e \in \ell_2(G)$. In this way one may view the elements of $C_r^*(G)$ as certain functions $f(e) = \sum_{g\in G}\langle f(e), g\rangle g$ which in particular satisfy $\sum_{g\in G} |\langle f(e), g\rangle|^2 < \infty$. One then defines $\kappa_{C_r^*} : C_r^*(G) \to \mathbb{C}$ by

$$\kappa_{C_r^*}(f) = \langle f(e), e\rangle$$

and checks that the adjoint f^* of f satisfies $\kappa_{C_r^*}(f^*) = \overline{\kappa_{C_r^*}(f)}$. Moreover, $\kappa_{C_r^*}$ has the trace property $\kappa_{C_r^*}(f_1 \circ f_2) = \kappa_{C_r^*}(f_2 \circ f_1)$. It follows that if $f \in C_r^*(G)$ is conjugate to a self-adjoint element (see for instance [19]), then $\kappa_{C_r^*}(f) \in \mathbb{R}$. Also, $\kappa_{C_r^*}(f^*f) = 0$ implies $f(e) = 0$, thus $f = 0$ and it follows that $\kappa_{C_r^*}$ is a *faithful* trace on $C_r^*(G)$. As a result, for an idempotent $\epsilon \in C_r^*(G)$ one has

$$0 \le \kappa_{C_r^*}(\epsilon) \le 1.$$

Indeed, using the fact that an idempotent in $C_r^*(G)$ is conjugate to a self-adjoint idempotent, one may assume that ϵ is self-adjoint so that

$$\kappa_{C_r^*}(\epsilon) = \langle \epsilon(e), e\rangle = \langle \epsilon^2(e), e\rangle = \langle \epsilon(e), \epsilon^*(e)\rangle = \langle \epsilon(e), \epsilon(e)\rangle \ge 0,$$

and similarly, $\kappa_{C_r^*}(1-\epsilon) \ge 0$ so that $\kappa_{C_r^*}(\epsilon) \le 1$. Moreover, the faithfulness of the trace implies that $\kappa_{C_r^*}(\epsilon) = 0$ only if $\epsilon = 0$ and, by considering $1-\epsilon$, that $\kappa_{C_r^*}(\epsilon) = 1$ implies $\epsilon = 1$. One extends $\kappa_{C_r^*}$ to a homomorphism $\kappa_{C_r^*} : K_0^{top}(C_r^*(G)) \to \mathbb{R}$ by defining it on an idempotent matrix $A = (a_{ij})$ in $C_r^*(G) \otimes_{\mathbb{C}} M_n(\mathbb{C})$ to be $\sum \kappa_{C_r^*}(a_{ii})$. As a result, if $\epsilon \in C_r^*(G)$ is an idempotent defining the element $[\epsilon] \in K_0^{top}(C_r^*(G))$, $0 \le \kappa_{C_r^*}([\epsilon]) \le 1$, and $\kappa_{C_r^*}([\epsilon]) = 0$ implies that $\epsilon = 0$ as well as $\kappa_{C_r^*}([\epsilon]) = 1$ implies $\epsilon = 1$. Therefore, if $\kappa_{C_r^*}$ takes integer values on $K_0^{top}(C_r^*(G))$, $C_r^*(G)$ has no non-trivial idempotent. Note that the argument also proves the classical result that for an arbitrary (not necessarily torsion-free) G, the integral group algebra $\mathbb{Z}[G]$ does not admit any non-trivial idempotent, since for an idempotent $\epsilon \in \mathbb{Z}[G]$ obviously $\kappa_{C_r^*}(\epsilon)$ is an integer. Of course, via $\mathbb{C}[G] \to C_r^*(G)$, $\kappa_{C_r^*}$ yields the *classical Kaplansky trace* $\kappa_{\mathbb{C}} : K_0^{alg}(\mathbb{C}[G]) \to \mathbb{R}$, which is known to take values in $\mathbb{Q}$ (Zalesskii's Theorem [19]).

We will consider the following two trace conjectures.

TC: Trace Conjecture

- *For a torsion-free group G the trace $\kappa_{\mathbb{C}} : K_0^{alg}(\mathbb{C}[G]) \to \mathbb{Q}$ assumes integer values only.*

Clearly **TC** $\Longrightarrow$ **IC**, since assuming **TC**, an idempotent $\epsilon \in \mathbb{Q}[G]$ gives rise to an idempotent $\epsilon \in C_r^*(G)$ with trace 0 or 1, thus ϵ equals 0 or 1.

STC: Strong Trace Conjecture

- *For a torsion-free group G the trace $\kappa_{C_r^*} : K_0^{top}(C_r^*(G)) \to \mathbb{R}$ assumes integer values only.*

These two conjectures are subgroup closed, as one sees from the natural commutative diagrams associated with $H < G$: for **TC** consider

$$\begin{array}{ccc} K_0^{alg}(\mathbb{C}[H]) & \xrightarrow{\mathrm{Ind}_H^G} & K_0^{alg}(\mathbb{C}[G]) \\ \kappa_{\mathbb{C}} \downarrow & & \downarrow \kappa_{\mathbb{C}} \\ \mathbb{Q} & = & \mathbb{Q} \end{array}$$

and similarly for the case of **STC**.

Obviously **STC** implies **TC**: one has a natural commutative diagram

$$\begin{array}{ccc} K_0^{alg}(\mathbb{C}[G]) & \xrightarrow{\kappa_{\mathbb{C}}} & \mathbb{Q} \\ \downarrow & & \downarrow \\ K_0^{top}(C_r^*(G)) & \xrightarrow{\kappa_{C_r^*}} & \mathbb{R} \end{array}$$

These conjectures are linked to **BCC** as follows. One considers the partial assembly map

$$A(G) : K_0(BG) = K_0^G(EG) \to K_0^G(\underline{E}G) \to K_0^{top}(C_r^*(G)),$$

induced by the G-maps $EG \to \underline{E}G \to \{e\}$. According to Atiyah's *ℓ_2-Index Theorem* [2], $\kappa_{C_r^*}$ takes integer values on the image of $A(G)$. For the proof one needs to identify our definitions of K^G-homology with an analytic version due to Kasparov, and similarly one needs to identify the associated assembly maps. We present here a direct proof of the integrality result, avoiding the use of the Atiyah ℓ_2-Index Theorem.

Lemma 7.1. *Let G be an arbitrary group and*

$$\kappa_{C_r^*} : K_0^{top}(C_r^*(G)) \to \mathbb{R}$$

the Kaplansky trace. Let

$$A(G) : K_0(BG) = K_0^G(EG) \to K_0^G(\underline{E}G) \to K_0^{top}(C_r^*(G))$$

be the partial assembly map induced by $EG \to \underline{E}G \to \{e\}$. Then

$$\kappa_{C_r^*}(\operatorname{Im} A(G)) = \mathbb{Z}.$$

Proof. Consider the embedding $G \rightarrowtail \beta(G)$ with $\beta(G)$ acyclic (cf. 5.24). Associated with the diagram
$$G \rightarrowtail \beta(G) \leftarrowtail \{e\}$$
we have a commutative diagram
$$\begin{array}{ccccc}
K_0(BG) & \xrightarrow{A(G)} & K_0^{top}(C_r^*(G)) & \xrightarrow{\kappa_{C_r^*}} & \mathbb{R} \\
\downarrow & & \downarrow & & \downarrow \\
K_0(B\beta(G)) & \xrightarrow{A(\beta(G))} & K_0^{top}(C_r^*(\beta(G))) & \xrightarrow{\kappa_{C_r^*}} & \mathbb{R} \\
\cong\uparrow & & \uparrow & & \uparrow \\
K_0(B\{e\}) & \xrightarrow[\cong]{A(\{e\})} & K_0(C_r^*(\{e\})) & \xrightarrow[\cong]{\kappa_{C_r^*}} & \mathbb{Z}
\end{array}$$
and the result follows. □

As we will show in [27], the techniques displayed in this proof can be used to give a purely algebraic proof of Atiyah's ℓ_2-Index Theorem.

Although G is not assumed to be torsion-free in the previous lemma, its use is essentially limited to that case. Namely, if G contains an element of order $n > 1$ then
$$\frac{1}{n} \in \operatorname{Im}(\kappa_{C_r^*} : K_0^{top}(C_r^*(G)) \to \mathbb{R}).$$
Indeed, if P denotes the projective $C_r^*(G)$-module induced up from the trivial one-dimensional $\mathbb{Z}/n\mathbb{Z}$-module $\mathbb{C}$ via $\mathbb{Z}/n\mathbb{Z} < G$ then
$$\kappa_{C_r^*}([P]) = \frac{1}{n}.$$
Better results on the image of $\kappa_{C_r^*}$ for groups with torsion are obtained by taking into account the contribution of the centralizers of elements of finite order; for the corresponding generalization of Lemma 7.1 see Lück [89] as well as [11].

Lemma 7.2. **BCC** $\Longrightarrow$ **STC** $\Longrightarrow$ **SIC** $\Longrightarrow$ **IC**

Proof. For G a torsion-free group, the assembly map
$$A(G) : K_0(BG) \to K_0^{top}(C_r^*(G))$$
is just the Baum–Connes assembly map. The surjectivity of $A(G)$ ensures by the previous Lemma that $\kappa_{C_r^*}$ is integer valued. We have already explained how **STC** implies the strong idempotent conjecture. □

Remark 7.3. One might be tempted to conjecture that, for an arbitrary group G, the trace $\kappa_{C_r^*}$ takes values in the subgroup of $\mathbb{Q}$ generated additively by the numbers of the form $1/n$, where $n = |F|$ for some finite subgroup $F < G$. But this is not so: Roy constructed in [114] a group G with 3-torsion but no element of order 9 and such that there exists an $x \in K_0^{top}(C_r^*(G))$ with $\kappa_{C_r^*}(x) \in \frac{1}{9}\mathbb{Z} \setminus \frac{1}{3}\mathbb{Z}$. The correct generalization is given by a theorem of Lück [89] which states that the Kaplansky trace $\kappa_{C_r^*}$ maps the image of the Baum–Connes assembly map into the *subring* of $\mathbb{Q}$ generated by $\{1/n \mid G$ has an element of order $n\}$. It is an open question whether the values of $\kappa_{C_r^*}$ are, in general, rational numbers.

Let M be a connected, closed and oriented smooth manifold. It has a *Hirzebruch* L-class $L^M = \sum L_{4i}^M \in H^*(M;\mathbb{Q})$, $L_{4i}^M \in H^{4i}(M;\mathbb{Q})$, which is a certain polynomial in the *Pontrjagin* classes of M. We write $L_M \in H_*(M;\mathbb{Q})$ for the Poincaré dual class $L^M \cap [M] = \sum L_M^{m-4i}$, with $m = \dim M$. The classical signature theorem, due to Hirzebruch, states that for $\dim M = 4m$, $L_{4m}^M \cap [M] \in H_0(M;\mathbb{Q}) = \mathbb{Q}$ is the signature of the cup-product quadratic form on $H^{2m}(M;\mathbb{R})$, which is obviously an oriented homotopy invariant. The *Novikov Conjecture* states that $f_*(L_M) \in H_*(BG;\mathbb{Q})$ is an oriented homotopy invariant, where $G = \pi_1(M)$ and $f : M \to BG$ classifies the universal cover of M. Equivalently, this can be expressed by stating that the *higher signatures* $\sigma_u(M) \in \mathbb{Q}$, defined by

$$\sigma_u(M) = \langle L^M \cup f^*(u), [M]\rangle$$

are, for all $u \in H^*(BG;\mathbb{Q})$, oriented homotopy invariants; note that the case of $u = 1 \in H^0(BG;\mathbb{Q})$ corresponds, when $\dim M$ is a multiple of 4, to the classical signature $\sigma(M)$ of M.

NC: Novikov Conjecture

- *The higher signatures of connected, closed and oriented smooth manifolds are oriented homotopy invariants.*

The *Strong Novikov Conjecture* for G asserts that the natural injection $K_*(BG) \otimes \mathbb{Q} \rightarrowtail K_*^G(\underline{E}G) \otimes \mathbb{Q}$ (cf. 6.2) followed by the rationalized Baum–Connes assembly map is injective.

SNC: Strong Novikov Conjecture

- *For any group G the natural map*

$$A : K_*(BG) \otimes \mathbb{Q} \longrightarrow K_*^{top}(C_r^*(G)) \otimes \mathbb{Q}$$

  induced by $EG \to \underline{E}G \to \{*\}$ is injective.

It is not known whether **NC** or **SNC** are subgroup closed.

Lemma 7.4. **BCC $\Longrightarrow$ SNC $\Longrightarrow$ NC**

Proof. The first implication is plain. For the second, one needs to know that the images of the Poincaré dual classes L_M^i in $K_i^{top}(C_r^*(G))$ via

$$H_i(M;\mathbb{Q}) \xrightarrow{f_*} H_i(BG;\mathbb{Q}) \rightarrowtail K_i(BG) \otimes \mathbb{Q} \longrightarrow K_i^{top}(C_r^*(G)) \otimes \mathbb{Q}$$

are oriented homotopy invariants. But these images may be interpreted as *analytic* signatures, which are known to be oriented homotopy invariants (for more details, see [4] and [99]). □

Remark 7.5. The main method used to prove injectivity of the assembly map $K_*^G(\underline{E}G) \to K_*^{top}(C_r^*(G))$ involves the *Dirac-dual Dirac* method (as described for instance in [130]). The **SNC** has been established for a variety of groups, including all groups satisfying both (RD) and (PC) , see [29]. Examples of such groups include all hyperbolic groups [55], [66] ; (RD) stands for the *Rapid Decay Property* and (PC) for *Polynomial Cohomology*, see Connes–Moscovici [29], where these terms are explained. For new results on groups satisfying (RD) see Chatterji [26], where she proves that cocompact lattices in a product involving rank 1 simple Lie groups, the groups $S\ell_3(\mathbb{R}), S\ell_3(\mathbb{C}), S\ell_3(\mathbb{H})$ and $E_{6(-26)}$, satisfy the Baum–Connes Conjecture (the case of cocompact lattices in $S\ell_3(\mathbb{R})$ and $S\ell_3(\mathbb{C})$ goes back to Lafforgue [75]). In a recent paper Mineyev and Yu proved the Baum–Connes Conjecture for hyperbolic groups and their subgroups [98]. Their work is based on a recent paper by Lafforgue [76], where the **BCC** is proved for *strongly bolic* groups with property (RD). The **SNC** on the other hand has also been proved for discrete subgroups of virtually connected Lie groups (cf. Kasparov [70]). More results can be found in Higson [58] as well as Higson and Roe [62].

Related to the strong Novikov conjecture is the *Gromov–Lawson–Rosenberg Conjecture*. Let M be a closed connected and oriented smooth manifold. Its $\hat{A}$-genus $\hat{A}^M \in H^*(M;\mathbb{Q})$ is a certain polynomial in the Pontrjagin classes of M. Let $\hat{A}_M \in H_*(M;\mathbb{Q})$ denote its Poincaré dual class. As before, denote by $f : M \to BG$ the map which classifies the universal cover of M. The vanishing of $f_*(\hat{A}_M) \in H_*(BG;\mathbb{Q})$ is equivalent to the vanishing of all the *higher $\hat{A}$-genera* $\hat{A}_u(M) \in \mathbb{Q}$, which are defined by

$$\hat{A}_u(M) = \langle \hat{A}^M \cup f^*(u), [M] \rangle,$$

where $u \in H^*(BG;\mathbb{Q})$. Note that for $u = 1 \in H^0(BG;\mathbb{Q})$

$$\hat{A}_1(M) = \langle \hat{A}^M, [M] \rangle =: \hat{A}(M),$$

the classical $\hat{A}$-genus of M. It is known (Lichnerowicz [78]) that for a spin manifold which admits a metric with positive scalar curvature, the $\hat{A}$-genus vanishes.

GLRC: Gromov–Lawson–Rosenberg Conjecture

- *Let M be a closed connected smooth spin manifold. Assume that M admits a Riemannian metric with positive scalar curvature. Then the higher $\hat{A}$-genera of M all vanish.*

We took our formulation of **GLRC** from [5]. The reader should be aware that in other places, by *Gromov–Lawson–Rosenberg Conjecture* something different might be meant. For recent results on that topic, one should consult Stolz's survey [127].

Lemma 7.6. **SNC $\Longrightarrow$ GLRC**

Proof. Let $f_*(\hat{A}_M) \in H_*(BG;\mathbb{Q})$ be as defined above. By a classical result of Lichnerowicz, for every i the i-component of $f_*(\hat{A}_M)$ under the natural map

$$H_i(BG;\mathbb{Q}) \xrightarrow{\text{mono}} K_i(BG) \otimes \mathbb{Q} \xrightarrow{A} K_i^{top}(C_r^*(G)) \otimes \mathbb{Q}$$

is zero. Thus, if A is injective, the claim follows (for more details see Rosenberg [112]). □

Remark 7.7. The converse of **GLRC** is false. Schick constructed in [117] a 5-dimensional closed spin manifold with vanishing higher $\hat{A}$-genera, which cannot carry any Riemannian metric with positive scalar curvature.

The *Zero-In-The-Spectrum Conjecture* goes back to Gromov, who asked whether for a closed, aspherical, connected and oriented Riemannian manifold M there always exists some $p \geq 0$, such that zero belongs to the spectrum of the Laplace–Beltrami operator Δ_p acting on square integrable p-forms on the universal cover $\tilde{M}$ of M. If $\pi = \pi_1(M)$ and $\ell_2(\pi)$ denotes the Hilbert π-module of square summable complex valued functions on π, then *zero not belonging to the spectrum of* $\Delta = \Delta_*$ can also be expressed as the vanishing of the (unreduced) ℓ_2-homology groups, $H_*^{(2)}(M;\ell_2(\pi))= H_*^{\pi}(\tilde{M};\ell_2(\pi))=0$, which are the homology groups of the complex $C_*(\tilde{M}) \otimes_\pi \ell_2(\pi)$. Equivalently, this can be expressed as the vanishing of the homology groups with (local) coefficients in the π-module $C_r^*(\pi)$, $H_*^{\pi}(\tilde{M};C_r^*(\pi)) = 0$ (cf. [64]), or, by the vanishing of $H_*^{\pi}(\tilde{M};\mathcal{N}(\pi))$, where $\mathcal{N}(\pi)$ denotes the von Neumann algebra of π (cf. [84]).

0ϵSC: Zero-In-The-Spectrum Conjecture

- *Let M be a closed, connected, oriented and aspherical Riemannian manifold with fundamental group π. Then, for some $i \geq 0$, $H_i^{\pi}(\tilde{M};C_r^*(\pi)) \neq 0$.*

For instance, one of the characteristic properties of amenable groups is the existence of a surjective map $C_r^*(\pi) \to \mathbb{C}$ of C^*-algebras, mapping each $g \in \pi$ to 1. Therefore for amenable π one has

$$H_0^{\pi}(\tilde{M};C_r^*(\pi)) \cong C_r^*(\pi) \otimes_{\mathbb{C}[\pi]} \mathbb{C} \neq 0,$$

proving the **0ϵSC** for such groups. Note that we did not need to assume here M to be aspherical (compare with Remark 7.9).

The following Theorem is due to Lott [82].

Theorem 7.8. **SNC $\Longrightarrow$ 0ϵSC**

Proof. We first consider the case of $\dim M$ even. The Poincaré dual class $L_M \in H_*(M;\mathbb{Q})=H_*(\pi;\mathbb{Q})$ of the Hirzebruch L-class of M is non-zero, because its top component, which is the Poincaré dual of $1 \in H^0(M;\mathbb{Q})$, equals the fundamental class $[M]$. Therefore, assuming **SNC**, its image $A(L_M)$ in $K_0(C_r^*(\pi)) \otimes \mathbb{Q}$ under the partial assembly map $A : H_{ev}(\pi;\mathbb{Q}) \to K_0(C_r^*(\pi)) \otimes \mathbb{Q}$ is non-zero. But by the *Higher Index Theorem* (cf. [82]), $A(L_M)$ equals the equivariant index of the operator $\tilde{d}+\tilde{d}^*$, which acts on $C_*(\tilde{M}) \otimes_\pi \ell_2(\pi)$. It follows that this operator is not invertible, and so neither is its square Δ. Therefore, 0 does belong to the spectrum of Δ. The case of $\dim M$ odd is dealt with by considering $M \times S^1$; for details the reader should consult [82]. □

Of course, the groups $H_i^\pi(\tilde{M}; C_r^*(\pi))$ depend on the homotopy type of M only. One could thus formulate a slightly stronger conjecture as follows.

S0ϵSC: Strong Zero-In-The-Spectrum Conjecture

- *Let X be a finite aspherical CW-complex with fundamental group π. Then, for some $i \geq 0$, $H_i^\pi(\tilde{X}; C_r^*(\pi)) \neq 0$.*

Remark 7.9. The condition that X should be aspherical cannot be dropped. There exists a finite connected CW-complex Y with fundamental group π isomorphic to $(\mathbb{Z} * \mathbb{Z})^3$ satisfying $H_*^\pi(\tilde{Y}; C_r^*(\pi)) = 0$ (cf. Farber–Weinberger [48] and Higson–Roe–Schick [64]). By embedding Y into some Euclidean space and taking for M the boundary of a regular neighborhood of Y one finds an example of a closed, connected manifold M with fundamental group π satisfying $H_*^\pi(\tilde{M}; C_r^*(\pi))=0$. Therefore, in **0ϵSC** the condition that M be aspherical cannot be dropped either. We don't know whether **BCC** implies **S0ϵSC**.

The group algebra $\mathbb{C}[\mathbb{Z}]$ has no non-trivial zero divisors. This can be seen by thinking of the elements of $\mathbb{C}[\mathbb{Z}]$ as *Laurent polynomials* and by concentrating on the leading term in a product of two Laurent polynomials. This argument can be generalized to left orderable groups $(G, \leq)$, see Rolfsen–Zhu [111]. Recall that a group G is called left orderable, if it admits a total order which is preserved by left translations. So, if G is left orderable, $\mathbb{C}[G]$ has no non-trivial zero divisors. Examples of left orderable groups include all torsion-free abelian groups, free groups, braid groups (cf. Rolfsen–Zhu [111] and Dehornoy [32]), the *Thompson group F* (cf. [20]) – to name a few. One can show that the class of left orderable groups is closed under directed unions, free products and extensions. There are more families of groups, for which it is known that $\mathbb{C}[G]$ does not admit any zero divisors $\neq 0$. For instance, using other techniques, Linnell proved in [80] that for a torsion-free G possessing a normal, free subgroup F with G/F elementary amenable, $\mathbb{C}[G]$ has no non-trivial zero-divisors.

On the other hand, the completion $C_r^*(\mathbb{Z})$ of $\mathbb{C}[\mathbb{Z}]$ has a lot of zero divisors, being isomorphic to $C(S^1)$: any two continuous functions $f, g : S^1 \to \mathbb{C}$ with disjoint supports satisfy $f(x) \cdot g(x) = 0$, $\forall x$, and define therefore zero divisors $f, g \in C(S^1)$. This implies that for any group $G \neq \{e\}$, $C_r^*(G)$ admits non-trivial zero divisors. Namely, if $x \in G$ is a torsion element, $1 - x \in C_r^*(G)$ is a zero divisor, and if $x \in G$ has infinite order, one has an injective map $C(S^1) \to C_r^*(G)$, displaying non-trivial zero divisors.

ZDC: Zero Divisor Conjecture

- *For G a torsion-free group, $\mathbb{Q}[G]$ has no non-trivial zero divisors*

Clearly, non-trivial idempotents in $\mathbb{Q}[G]$ are also non-trivial zero divisors so that the following implication holds.

Lemma 7.10. **ZDC $\Longrightarrow$ IC**

Of course, we could formulate the **ZDC** for $\mathbb{C}[G]$ instead of $\mathbb{Q}[G]$. We have chosen $\mathbb{Q}[G]$, because there is in this case a close relationship with *Atiyah's Conjecture* on the rationality of ℓ_2-Betti numbers (cf. Lemma 7.12). For background on ℓ_2-(co)homology see Eckmann's survey [42] or Lück's comprehensive treatment in [85], [86] [90] as well as Lück's forthcoming book [91]. If M is a connected smooth manifold with a cocompact proper and free action of a discrete group G, the ℓ_2-Betti numbers $\beta_i(M,G)$ are defined to be the von Neumann dimensions of the reduced ℓ_2-homology groups $\overline{H}_i^{(2)}(M;\ell_2(G))$, which are Hilbert G-modules of finite type. For instance, if G is finite, $\beta_0(M,G) = 1/|G|$. Atiyah asked in [2] whether these ℓ_2-Betti numbers, which are a priori real numbers, are rational. For G torsion-free, they might even be integers. Instead of assuming M to be a manifold as Atiyah did in his original approach, one could equivalently just look at a free, cocompact G-CW-complex X and ask whether the ℓ_2-Betti numbers $\beta_i(X,G)$ are rational. (Again, these Betti numbers are defined as the von Neumann dimensions of the reduced ℓ_2-homology groups of X with respect to the action of G. If (X,G) is given and $i \geq 0$ fixed, one can construct a smooth cocompact G-manifold (M,G) with $\beta_i(M,G) = \beta_i(X,G)$ for that particular i as follows. One embeds X/G into a closed manifold N such that $X/G \subset N$ is $(i+2)$-connected. Then one takes for M the covering space of N associated to the obvious surjection $\pi_1(N) = \pi_1(X) \to G$.)

AC: Atiyah Conjecture

- *Let X be a free, cocompact G-CW-complex. Then the ℓ_2-Betti numbers $\beta_i(X,G)$ are all rational. In case G is torsion-free, they are integers.*

Linnell proved **AC** for groups G with *bounded* torsion subgroups, such that G fits in a short exact sequence $F \to G \to G/F$ with F a free group and G/F elementary amenable [80]. For other results, see Schick [120] as well as the recent preprint [36] by Dodziuk, Linnell, Mathai, Schick and Yates. Note also that **AC** is subgroup closed: if G satisfies **AC** and Y is a cocompact H-CW-complex for some $H < G$, then

$$\beta_i(Y,H) = \beta_i(G \times_H Y, G)$$

because the von Neumann dimension over $\mathcal{N}(H)$ of a finitely generated projective $\mathcal{N}(H)$-module M is the same as the von Neumann dimension of the induced module $\mathcal{N}(G) \otimes_{\mathcal{N}(H)} M$ over $\mathcal{N}(G)$.

Remark 7.11. One could, in case G has torsion, think that $\beta_i(X,G)$ lies in the subgroup of $\mathbb{Q}$ additively generated by the elements $1/|H|$, where $H < G$ runs over the finite subgroups of G. But this turns out to be false; in [51] one finds a closed 7-manifold M with fundamental group G without 3-torsion and such that its universal cover $\tilde{M}$ satisfies $\beta_3(\tilde{M},G) = 1/3$. In view of recent work by Dicks and Schick [33] it even seems likely that, in case G has torsion, there are examples of cocompact free G-CW-complexes with irrational ℓ_2-Betti numbers.

Lemma 7.12. **AC $\Longrightarrow$ ZDC**

Proof. We sketch the argument from [42]. Let K be a torsion-free group. If $s \in \mathbb{Q}[K]$ is a non-trivial zero divisor, then s is also a non-trivial zero divisor in $\mathbb{Q}[G]$ for some finitely generated subgroup $G < K$. We can even assume that $s \in \mathbb{Z}[G]$, by replacing s by some integer multiple, if necessary. Write $\ker(s) \subset \mathbb{Z}[G]$ for the left G-module given as the kernel of the right multiplication by s on $\mathbb{Z}[G]$. One can then construct a 2-dimensional connected free G-CW-complex X with X/G finite and $H_2(X;\mathbb{Z}) = \ker(s)$. Now $\beta_2(X,G)$ is the von Neumann dimension of the kernel of the right multiplication by s on $\ell_2(G)$, which is a certain Hilbert submodule $T \subset \ell_2(G)$. Since, by our assumption, $T \neq 0$ and G is torsion-free, it follows that the von Neumann dimension of T must equal 1 by **AC**. But this implies that $T = \ell_2(G)$, from which we conclude $s = 0$, contrary to our assumption. □

A closely related conjecture is the following *Embedding Conjecture*. For this, we need to reformulate the Atiyah Conjecture in a slightly stronger, algebraic setting.

SAC: Strong Atiyah Conjecture

- *Let A be a complex $n \times m$-matrix, defining a bounded G-invariant operator $\ell_2^n(G) \to \ell_2^m(G)$ via right-multiplication with A. Then the von Neumann dimension of the closure of the image $\ell_2(G)^n \cdot A$ in $\ell_2^m(G)$ is a rational number; it is an integer in case G is torsion-free.*

We haven't defined the von Neumann dimension of a Hilbert G-module; the interested reader should consult the many references we cited earlier. We won't discuss therefore the easy argument showing that **SAC** implies **AC**. It is moreover easy to see that **SAC** implies **TC**; from the definitions it is plain that for a finitely generated projective $\mathbb{C}[G]$-module P, given as the image under right-multiplication by the idempotent complex $n \times n$-matrix $A, P = \mathbb{C}[G]^n \cdot A \subset \mathbb{C}[G]^n$, one finds that $\kappa_{\mathbb{C}}(P)$ equals the von Neumann dimension of the closure of $\ell_2(G)^n \cdot A \subset \ell_2(G)^n$. Our reason for mentioning the Strong Atiyah Conjecture is its connection with the *Embedding Conjecture.*

EC: Embedding Conjecture

- *If G is torsion-free, then $\mathbb{C}[G]$ admits an embedding into a skew field.*

Linnell proved in [80] that **SAC** ⇒ **EC** by constructing for a torsion-free group G satisfying **AC** an explicit embedding $\mathbb{C}[G] \subset D(G)$, with $D(G)$ a skew field. His $D(G)$ is the division closure of $\mathbb{C}[G]$ in the algebra $U(G)$ of densely defined operators $\ell_2(G) \to \ell_2(G)$, affiliated to the von Neumann algebra $\mathcal{N}(G)$. Of course **EC** ⇒ **ZDC** so that Linnell's result provides an other proof that **AC** ⇒ **ZDC**.

The Hattori–Stallings trace considered earlier in the context of group representations can be viewed in a more general setting as follows.

Let S be an R-algebra and M an R-module. An R-linear map $f : S \to M$ is called a *trace*, if $f(ab) = f(ba)$ for all $a, b \in S$. There is a *universal* trace

$$T_{uni} : S \to S/[S,S],$$

where $[S,S] \subset S$ denotes the R-module generated by the elements of the form $ab - ba$ in S; every trace $S \to M$ factors uniquely through T_{uni}. We are mainly interested in the case of $S = R[\pi]$, the group R-algebra of the group π, R a commutative ring with unit. We write $\mathrm{CoC}[\pi]$ for the free abelian group generated by the conjugacy classes of the group π. Therefore, $\mathrm{CoC}[\pi] \otimes R$ is the free R-module of R-valued class functions on π with finite support; in particular for a finite group π, $\mathrm{CoC}[\pi] \otimes \mathbb{C}$ is, via the character of a representation, naturally isomorphic to the underlying abelian group of the complexified complex representation ring $R_{\mathbb{C}}(\pi) \otimes \mathbb{C}$. In view of the following lemma, the universal trace for group algebras can be considered as a generalization of the character map for group representations.

Lemma 7.13. *If $S = R[\pi]$, then $S/[S,S]$ is naturally isomorphic to* $\mathrm{CoC}[\pi] \otimes R$, *where* $\mathrm{CoC}[\pi]$ *denotes the free abelian group generated by the conjugacy classes of* π.

Indeed, since the R-submodule of $R[\pi]$ generated by the elements $xy - yx$ with $x, y \in \pi$ is the same as the one generated by the elements of the form $u - vuv^{-1}$ with $u, v \in \pi$, the claim of the lemma follows. Note also that for a finite group π, $\mathrm{CoC}[\pi] \otimes \mathbb{C}$ is just what we earlier (cf. page 17) denoted by $\mathbb{C}[\mathrm{FC}(\pi)]$.

Any trace $t : S \to M$ gives rise to a trace $M_n(t) : M_n(S) \to M$ by putting $M_n(t)(A) = \sum_i t(a_{ii})$, where $A = (a_{ij}) \in M_n(S)$. One can then define the trace $t(\phi)$ for any S-map $\phi : F \to F$, F a finitely generated free S-module by choosing an S-basis of F and putting $t(\phi) = M_n(t)(A(\phi))$, where $A(\phi)$ is the matrix of ϕ with respect to the chosen basis. Because of the trace property of t, $t(\phi)$ is well-defined. To define the trace of a finitely generated projective S-module P one represents P as the image of an idempotent S-map $\phi_P : S^n \to S^n$ and one puts $t(P) := t(\phi_P)$; it only depends on the isomorphism class of P. One therefore obtains a homomorphism of abelian groups $t : K_0^{alg}(S) \to M$, where $K_0^{alg}(S)$ stands for the Grothendieck group of finitely generated projective S-modules.

In case of $t = T_{uni}$, the universal trace, this map is sometimes called the *Hattori–Stallings trace* [57]. We are particularly interested in the case $S = R[\pi]$, in which we denote the Hattori–Stallings trace by

$$HS_R : K_0^{alg}(R[\pi]) \to \mathrm{CoC}[\pi] \otimes R = R \oplus (\ ?\ ).$$

Note that $\mathrm{CoC}[\pi] \otimes R$ splits naturally into $R \oplus (\ ?\ )$, where the summand R corresponds to the conjugacy class of $e \in \pi$. The trace corresponding to that summand R is called the *Kaplansky* trace and we denote it by

$$\kappa_R^\pi : K_0^{alg}(R[\pi]) \to R.$$

We just write κ_R for κ_R^π, if π is clear from the context. As remarked earlier, for $R = \mathbb{C}$ the Kaplansky trace $\kappa_{\mathbb{C}}$ takes values in $\mathbb{Q}$ (Zalesskii's Theorem, see [19]). In general, the Hattori–Stallings trace $HS_{\mathbb{C}}$ takes values in $\mathrm{CoC}[\pi] \otimes \overline{\mathbb{Q}}$ (cf. Bass [3]).

Another basic trace is given by the augmentation $\epsilon : R[\pi] \to R$, which satisfies $\epsilon(g) = 1$ for all $g \in \pi$. It gives rise to

$$\epsilon_R : K_0^{alg}(R[\pi]) \to R.$$

For R a field of characteristic 0 this trace takes values in $\mathbb{Z} \cong \mathbb{Z} \cdot 1_R \subset R$, because for P a finitely generated projective $R[\pi]$-module, $\epsilon_R([P])$ corresponds then to the dimension of $R \otimes_{R[\pi]} P$ over R, considered as an element of $\mathbb{Z} \subset R$. Note also that $\epsilon_{\mathbb{C}} = \bar{\epsilon}_{\mathbb{C}} \circ HS_{\mathbb{C}}$, where $\bar{\epsilon}_{\mathbb{C}}$ denotes the map $\mathrm{CoC}[\pi] \otimes \mathbb{C} \to \mathbb{C}$ induced by $\epsilon_{\mathbb{C}}$.

The relationship between these traces is expressed by the *Bass Conjecture*, which comes in many variations depending on the ground ring R (cf. Bass [3]). We are interested here in the case $R = \mathbb{C}$.

BTC: Bass Trace Conjecture

- *The Hattori–Stallings trace*

$$HS_{\mathbb{C}} : K_0^{alg}(\mathbb{C}[G]) \to \mathrm{CoC}[G] \otimes \mathbb{C}$$

takes its values in $\mathbb{C}[FC(G)]$, the $\mathbb{C}$-vector space spanned by the conjugacy classes of elements of finite order.

It is not known whether **BTC** is subgroup closed. Note that for G torsion-free, **BTC** implies that $HS_{\mathbb{C}} = \kappa_{\mathbb{C}} = \epsilon_{\mathbb{C}}$ so that $\kappa_{\mathbb{C}}$ is integer valued: **BTC** $\Longrightarrow$ **TC**. Moreover, since $HS_{\mathbb{Z}} : K_0^{alg}(\mathbb{Z}[G]) \to \mathrm{CoC}[G]$ has all components corresponding to elements $g \neq e$ of finite order equal to zero (cf. Linnell [79, Lemma 4.1]), the **BTC** implies the classical version of the Bass conjecture, which states that

$$HS_{\mathbb{Z}} = \kappa_{\mathbb{Z}} = \epsilon_{\mathbb{Z}}.$$

In case G is torsion-free, **BTC** obviously implies that $\kappa_{\mathbb{C}} = \epsilon_{\mathbb{C}}$, which is known as the *Weak Bass Conjecture* for the case $R = \mathbb{C}$.

WBTC: Weak Bass Trace Conjecture

- *Let G be a torsion-free group. Then*

$$\kappa_{\mathbb{C}} = \epsilon_{\mathbb{C}} : K_0^{alg}(\mathbb{C}[G]) \to \mathbb{R}.$$

Of course, **WBTC** $\Longrightarrow$ **TC**. Note that the **WBTC** is subgroup closed. It is also plain from the definitions that if G is a directed union of subgroups G_α each of which satisfies **WBTC** then so does G. For the proof of **WBTC** for a particular group G it can be useful to pass to the maximal C^*-algebra $C^*(G)$ of G as follows. The augmentation $\mathbb{C}[G] \to \mathbb{C}$, given by mapping each group element to $1 \in \mathbb{C}$, extends to a C^*-algebra map $C^*(G) \to \mathbb{C}$, which can be thought of as the representation induced by $G \to \{e\}$ on the level of the maximal C^*-algebras. The resulting induced map of K-groups will be denoted by

$$\epsilon_{C^*}^G : K_0^{top}(C^*(G)) \to K_0^{top}(\mathbb{C}) = \mathbb{Z} \subset \mathbb{R};$$

we sometimes just write ϵ_{C^*} for this map, if G is clear from the context. Note that we have a commutative diagram

$$\begin{array}{ccccc} K_0(\mathbb{Z}[G]) & \longrightarrow & K_0(\mathbb{C}[G]) & \longrightarrow & K_0(C^*(G)) \\ \downarrow{\scriptstyle \epsilon_{\mathbb{Z}}} & & \downarrow{\scriptstyle \epsilon_{\mathbb{C}}} & & \downarrow{\scriptstyle \epsilon_{C^*}} \\ \mathbb{Z} & = & \mathbb{Z} & = & \mathbb{Z}. \end{array}$$

The point here is that we use $C^*(G)$ rather than $C_r^*(G)$, because the augmentation $\epsilon_{\mathbb{C}} : \mathbb{C}[G] \to \mathbb{C}$ does not, in general, admit an extension to $C_r^*(G)$. As a matter of fact, it is known that the augmentation $\epsilon : \mathbb{C}[G] \to \mathbb{C}$ extends to a map of C^*-algebras $C_r^*(G) \to \mathbb{C}$ if and only if G is amenable – a proof of this is presented in [62].

We define the *Kaplansky* trace

$$\kappa_{C^*} = \kappa_{C^*}^G : K_0(C^*(G)) \to \mathbb{R}$$

to be the composite

$$\kappa_{C_r^*} \circ \pi_* : K_0(C^*(G)) \to K_0(C_r^*(G)) \to \mathbb{R},$$

where π_* is the map induced by the projection $C^*(G) \twoheadrightarrow C_r^*(G)$. As a result, there is a commutative diagram

$$\begin{array}{ccccccc}
K_0(\mathbb{Z}[G]) & \longrightarrow & K_0(\mathbb{C}[G]) & \longrightarrow & K_0(C^*(G)) & \longrightarrow & K_0(C_r^*(G)) \\
\downarrow{\scriptstyle \kappa_{\mathbb{Z}}} & & \downarrow{\scriptstyle \kappa_{\mathbb{C}}} & & \downarrow{\scriptstyle \kappa_{C^*}} & & \downarrow{\scriptstyle \kappa_{C_r^*}} \\
\mathbb{Z} & \longrightarrow & \mathbb{Q} & \longrightarrow & \mathbb{R} & = & \mathbb{R}.
\end{array}$$

Obviously, for an *injective* group homomorphism $G \to H$ one has a commutative diagram

$$\begin{array}{ccc}
K_0(C^*(G)) & \longrightarrow & K_0(C^*(H)) \\
{\scriptstyle \kappa_{C^*}^G}\downarrow & & \downarrow{\scriptstyle \kappa_{C^*}^H} \\
\mathbb{R} & = & \mathbb{R}.
\end{array}$$

The weak Bass conjecture is related to the Baum–Connes Conjecture as follows.

Lemma 7.14. *If G is a torsion-free, countable and K_*-amenable group which satisfies the* **BCC** *then G satisfies the* **WBTC**.

Proof. We consider the embedding $G \rightarrowtail \beta(G)$ defined in 5.24, with $\beta(G)$ K-amenable and acyclic. The diagram $G \rightarrowtail \beta(G) \leftarrowtail \{e\}$ then gives rise to a commutative diagram

$$\begin{array}{ccccccc}
K_0(BG) & \xrightarrow{\cong} & K_0(C_r^*(G)) & \xleftarrow{\cong} & K_0(C^*(G)) & \xrightarrow{\epsilon_{C^*}^G,\, \kappa_{C^*}^G} & \mathbb{R} \\
\downarrow & & & & {\scriptstyle \phi}\downarrow & & \| \\
K_0(B\beta(G)) & \longrightarrow & K_0(C_r^*(\beta(G))) & \xleftarrow{\cong} & K_0(C^*(\beta(G))) & \xrightarrow{\epsilon_{C^*}^{\beta G},\, \kappa_{C^*}^{\beta G}} & \mathbb{R} \\
{\scriptstyle \cong}\uparrow & & & & {\scriptstyle \psi}\uparrow & & \\
K_0(B\{e\}) & \xrightarrow{\cong} & K_0(\mathbb{C}) & \xleftarrow{\cong} & K_0(\mathbb{C}) & &
\end{array}$$

and we conclude that $\epsilon_{C^*}^{\beta G}, \kappa_{C^*}^{\beta G} : K_0(C^*(\beta(G))) \to \mathbb{R}$ agree on $\operatorname{Im}\phi$ (the diagram shows that $\operatorname{Im}\phi$ is the subgroup of $K_0(C^*(\beta(G)))$ generated by the stably free $C^*(\beta(G))$-modules). We conclude that

$$\epsilon_{C^*}^G = \kappa_{C^*}^G : K_0(C^*(G)) \to \mathbb{R}$$

which, by the commutativity of

$$\begin{array}{ccc} K_0(\mathbb{C}[G]) & \xrightarrow{\epsilon_{\mathbb{C}}^G,\, \kappa_{\mathbb{C}}^G} & \mathbb{R} \\ \downarrow & & \| \\ K_0(C^*(G)) & \xrightarrow{\epsilon_{C^*}^G = \kappa_{C^*}^G} & \mathbb{R} \end{array}$$

concludes the proof. □

This leads now to the following relationship between **WBTC** and **BCC**.

Theorem 7.15. *Let G be a torsion-free group belonging to the class* **LH$\mathcal{TH}$** *cf. 5.18). Then G satisfies the* **WBTC**. *In particular all torsion-free amenable groups satisfy the* **WBTC**.

Proof. Let G be the directed union of its countable subgroups $\{G_\alpha\}$. Then each G_α is torsion-free, K-amenable and satisfies **BCC** (cf. Theorem 5.18). The previous Lemma shows that each G_α satisfies **WBTC**. But this implies that G satisfies **WBTC** too. Indeed, if P is a finitely generated projective $\mathbb{C}[G]$-module, then P is isomorphic to the image under right multiplication

$$\mathbb{C}[G]^n \to \mathbb{C}[G]^n, \quad x \mapsto xA$$

for some idempotent $n \times n$-matrix A with entries in $\mathbb{C}[G]$. Clearly, A has entries in $\mathbb{C}[G_\alpha]^n$ for some countable subgroup G_α of G. But this implies that P is of the form $\mathbb{C}[G] \otimes_{G_\alpha} Q$ with Q a finitely generated projective $\mathbb{C}[G_\alpha]$-module. The result now follows, since

$$\epsilon_{\mathbb{C}}^G(P) = \epsilon_{\mathbb{C}}^{G_\alpha}(Q) = \kappa_{\mathbb{C}}^{G_\alpha}(Q) = \kappa_{\mathbb{C}}^G(P).$$

□

The full **BTC** is known for *elementary* amenable groups by a recent result of Farrell and Linnell [50]. In [11], a proof of **BTC** is given for *arbitrary* amenable groups. More generally, an ℓ_1-version of the Bass Conjecture is there proved for all groups whose countable subgroups have the Haagerup Property.

In dealing with the general **BTC**, we record the following basic and very useful result, due to Bass (cf. Proposition 9.2 of [3]).

Proposition 7.16. *If G is a counterexample to* **BTC** *then there exists $s \in G$ of infinite order and $N \geq 1$ such that for infinitely many primes p the element s is conjugate to s^{p^N}.*

As an application, we show that the *Mapping Class Groups* Γ_g satisfy the **BTC** (we don't know whether Γ_g satisfies **BCC**). Recall that Γ_g is the group of isotopy classes of orientation preserving diffeomorphisms of a closed smooth surface S_g of genus g (for background material on mapping class groups, the reader can consult [100]).

Corollary 7.17. *The mapping class groups* Γ_g *satisfy the* **BTC**.

Proof. According to Birman, Lubotzky and McCarthy [14] the solvable subgroups of Γ_g are all virtually abelian. Suppose that Γ_g contains a subgroup U of the form $\langle x, y | yxy^{-1} = x^n \rangle$ with $|n| > 1$, which is an extension of an abelian group by an abelian group, thus solvable. Suppose $A < U$ is an abelian subgroup of finite index in U. Then U contains for some $k, \ell > 0$ the non-commuting elements x^k and y^ℓ, which yields a contradiction. Therefore Γ_g does not have a subgroup of the form U and by 7.16 we conclude that Γ_g satisfies the **BTC**. □

Here is a short (and incomplete) list of groups for which the **BTC** has been verified.

- linear groups (i.e. groups which admit an embedding into $Gl_n(K)$ for some field K). This is already proved in Bass' paper [3] in case the group in question is torsion-free; for the general case, see Linnell's paper [79, Lemma 4.1]. In particular this applies for instance to the *Artin Braid Groups* B_n, since these are now known to be linear (cf. Bigelow [13] and Krammer [71])
- groups with $cd_{\mathbb{Q}} \leq 2$ (cf. Eckmann [41]; the argument presented there is for the case of $HS_{\mathbb{Q}}$, but it works the same way for $HS_{\mathbb{C}}$)
- semi-hyperbolic groups (which include hyperbolic groups as well as cocompact CAT(0)-groups; compare Eckmann [43]); this follows, using 7.16, from the fact that semi-hyperbolic groups do not admit subgroups of the form $\langle x, y | yxy^{-1} = x^n \rangle$ with $|n| > 1$ (cf. Corollary 4.19 of Bridson–Haefliger [17])
- elementary amenable groups (cf. Farrell–Linnell [50])
- amenable groups and, more generally, groups whose countable subgroups have the Haagerup Property [11].

For other results on the Bass Conjecture see for instance [43], [45], [115] and [119]. For its relationship with the **EC**, see Schafer's paper [116].

If $H < G$, there is a natural induction map

$$\mathrm{Ind}_H^G : K_0^{alg}(\mathbb{C}[H]) \to K_0^{alg}(\mathbb{C}[G])$$

given by the map $P \mapsto \mathbb{C}[G] \otimes_{\mathbb{C}[H]} P$ on finitely generated projective $\mathbb{C}[H]$-modules P. Since inner automorphisms of G induce the identity on $K_0^{alg}(\mathbb{C}[G])$, these induction maps fit together to define a map $\iota_G : R_{\mathbb{C}}^{\mathfrak{Fin}}(G) = \mathrm{colim}_{G/H \in \mathcal{O}_{\mathfrak{Fin}}(G)} R_{\mathbb{C}}(H) \to K_0^{alg}(\mathbb{C}[G])$. The following is an old conjecture.

PCGC: Projective Class Group Conjecture

- *For every group G the natural map*

$$\iota_G : \operatorname{colim}_{G/H \in \mathfrak{O}_{\mathfrak{Fin}}(G)} K_0^{alg}(\mathbb{C}[H]) = R_{\mathbb{C}}^{\mathfrak{Fin}}(G) \longrightarrow K_0^{alg}(\mathbb{C}[G])$$

is an isomorphism.

The **PCGC** obviously holds for finite groups. In case G is torsion-free, $R_{\mathbb{C}}^{\mathfrak{Fin}}(G) = \mathbb{Z}$ and $\iota_G(n)$ is the class of $\mathbb{C}[G]^n$ in $K_0^{alg}(\mathbb{C}[G])$ so that **PCGC** is, in this case, equivalent to the conjecture that all finitely generated projective $\mathbb{C}[G]$-modules are stably free, a statement which is known to be true for many classes of torsion-free groups (e.g. free groups and poly-$\mathbb{Z}$-groups). Also, since obviously for a general G, a finite subgroup $H < G$ and a finitely generated projective $\mathbb{C}[H]$-module P, $HS_{\mathbb{C}}([\mathbb{C}[G] \otimes_{\mathbb{C}[H]} P]) \in \mathbb{C}[\mathrm{FC}(G)]$, we conclude the following.

Lemma 7.18. **PCGC $\Longrightarrow$ BTC**

Note that there is a commutative diagram

$$\begin{array}{ccc} R_{\mathbb{C}}^{\mathfrak{Fin}}(G) & \longrightarrow & K_0^{alg}(\mathbb{C}[G]) \\ {\scriptstyle HS_G^{\mathfrak{Fin}}}\downarrow & & \downarrow{\scriptstyle HS_{\mathbb{C}}} \\ \mathbb{C}[\mathrm{FC}(G)] & \xrightarrow{mono} & \mathrm{CoC}(G) \otimes \mathbb{C}. \end{array}$$

The map $HS_G^{\mathfrak{Fin}}$ has been defined earlier (cf. Theorem 3.19), where it was shown that the induced map

$$HS_{G.\mathbb{C}}^{\mathfrak{Fin}} : R_{\mathbb{C}}^{\mathfrak{Fin}}(G) \otimes \mathbb{C} \to \mathbb{C}[\mathrm{FC}(G)]$$

is an isomorphism. This proves that **PCGC** implies the following *strong* version of the Bass Trace Conjecture.

SBTC: Strong Bass Trace Conjecture

- *For any group G the Hattori–Stallings trace induces an isomorphism*

$$K_0^{alg}(\mathbb{C}[G]) \otimes \mathbb{C} \xrightarrow{\cong} \mathbb{C}[\mathrm{FC}(G)],$$

where $\mathbb{C}[\mathrm{FC}(G)]$ denotes the $\mathbb{C}$-vector space spanned by the conjugacy classes of elements of finite order in G.

The following diagram sums up the relationship between the conjectures we described.

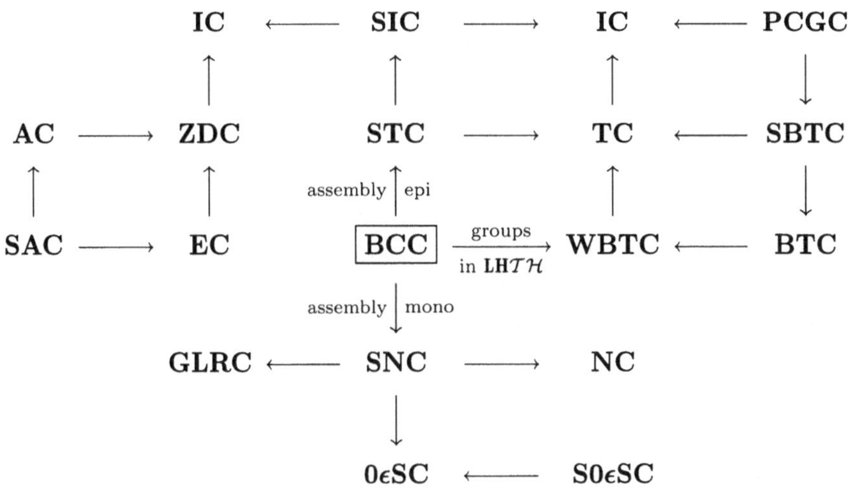
IC
SIC
IC
PCGC
AC
ZDC
STC
TC
SBTC
assembly
epi
SAC
EC
BCC
groups
in LHTH
WBTC
BTC
assembly
mono
GLRC
SNC
NC
0εSC
S0εSC

References

[1] M. A. Armstrong, *On the fundamental group of an orbit space*, Proc. Cambridge Philos. Soc. **61** (1965), 639–646.

[2] M. F. Atiyah, *Elliptic operators, discrete groups and von Neumann algebras,* Astérisque **32-3** (1976), 43–72.

[3] H. Bass, *Euler characteristics and characters of discrete groups*, Invent. Math. **35** (1976), 155–196.

[4] P. Baum and A. Connes, *Geometric K-theory for Lie groups and foliations,* L'Enseign. Math. (2) **46** (2000), no. 1–2, 3–42.

[5] P. Baum, A. Connes and N. Higson, *Classifying spaces for proper actions and K-theory of group C^*-algebras,* Contemporary Mathematics **167** (1994), 241–291.

[6] P. Baum and R. Douglas, *K homology and index theory,* Proc. Sympos. Pure Math. **38**, Part 1, Amer. Math. Soc. 1982, 117–173.

[7] G. Baumslag, E. Dyer and H. Heller, *The topology of a discrete group,* J. Pure Appl. Algebra **16** (1980) 1, 1–47.

[8] C. Béguin, H. Bettaieb and A. Valette, *K-theory for C^*-Algebras of One-Relator Groups,* K-Theory **16** (1999), 277–298.

[9] M. Bekka, P.-A. Cherix and A. Valette, *Proper affine isometric actions of amenable groups,* in: Novikov conjecture, index theorems and rigidity, London Math. Soc. Lecture Notes series **227**, Cambridge Univ. Press, 1995, vol. 2, 1–4.

[10] A. J. Berrick, *Universal groups, binate groups and acyclicity,* Group Theory (Singapore, 1987), de Gruyter, Berlin 1989, 253–266.

[11] A. J. Berrick, I. Chatterji and G. Mislin, *From acyclic groups to the Bass Conjecture for amenable groups*, preprint 2001.

[12] R. Bieri, *Homological Dimension of Discrete Groups,* Mathematics Notes, Queen Mary College, London, 2nd ed. 1981.

[13] S. Bigelow, *Braid groups are linear*, J. Amer. Math. Soc. (electronic) **14** (2001), 471–486.

[14] J. Birman, A. Lubotzky and J. McCarthy, *Abelian and solvable subgroups of the mapping class group,* Duke Math J. **50** (1983), 1107–1120.

[15] M. Bozejko, T. Januszkiewicz and R. Spatzier, *Infinite Coxeter groups do not have Kazhdan's property (T),* J. Operator Theory **19** (1988), 63–67.

[16] G. Bredon, *Equivariant Cohomology Theories,* Springer Lecture Notes in Math. **34**, 1967.

[17] M. R. Bridson and A. Haefliger, *Metric Spaces of Non-Positive Curvature,* Springer, Grundlehren **319**, 1999.

[18] K. S. Brown, *Cohomology of Groups,* Springer Graduate Text in Mathematics **87**, 1982.

[19] M. Burger and A. Valette, *Idempotents in complex group rings: theorems of Zalesskii and Bass revisited*, Journal of Lie Theory **8** (1998), 219–228.

[20] J. W. Cannon, W. J. Floyd and W. R. Parry, *Introductory notes on Richard Thompson's groups,* L'Enseignement Mathématique **42**, 1996, 215–256.

[21] G. Carlsson, E. K. Pedersen and J. Roe, *Controlled C^*-theory and the injectivity of the Baum-Connes map,* in preparation.

[22] J. Chabert, *Baum–Connes conjecture for some semi-direct products,* J. reine angew. Math. **521** (2000), 161–184.

[23] J. Chabert, *Stabilité de la conjecture de Baum–Connes pour certains produits semi-directs de groupes,* C. R. Acad. Sci. Paris, t. 328, Série I (1999), 1120–1132.

[24] J. Chabert and S. Echterhoff, *Permanence properties of the Baum–Connes conjecture,* Documenta Math. **6** (2001), 127–183.

[25] J. Chabert, S. Echterhoff and R. Meyer, *Deux remarques sur l'application de Baum–Connes,* C. R. Acad. Sci. Paris, t. **332**, Série I (2001), 607–610.

[26] I. Chatterji, *Property (RD) for cocompact lattices in a finite product of rank one Lie groups with some rank two Lie groups,* to appear in Geom. Dedicata.

[27] I. Chatterji and G. Mislin, *Atiyah's L^2-Index Theorem,* to appear in L'Enseign. Math.

[28] P.-A. Cherix, M. Cowling, P. Jolissaint, P. Julg and A. Valette, *Groups with the Haagerup property (Gromov's a-T-menability),* Birkhäuser, Progress in Math. **197**, 2001.

[29] A. Connes and H. Moscovici, *Cyclic cohomology, the Novikov conjecture and hyperbolic groups,* Topology **39** (1990), 345–388.

[30] J. Cuntz, *K-theoretic amenability for discrete groups,* J. reine angew. Math. **344** (1983), 180–195.

[31] J. F. Davis and W. Lück, *Spaces over a Category and Assembly Maps in Isomorphism Conjectures in K- and L-theory,* K-Theory **15** (1998), 201–252.

[32] P. Dehornoy, *Braid groups and left distributive operations,* Trans. Amer. Math. Soc. **345** (1994), no.1, 115–150.

[33] W. Dicks and T. Schick, *The spectral measure of certain elements of the complex group ring of a wreath product,* Geometriae Dedicata **93** (2002), 121–137.

[34] T. tom Dieck, *Orbittypen und äquivariante Homologie I,* Arch. Math. **23** (1972), 307–317.

[35] T. tom Dieck, *Transformation Groups*, Studies in Math. **8**, de Gruyter 1987.

[36] J. Dodziuk, P. Linnell, V. Mathai, T. Schick and S. Yates, *Approximating L^2-invariants, and the Atiyah conjecture,* preprint 2001.

[37] A. Dold, *Relations between ordinary and extraordinary homology*, mimeographed notes of the Colloquium on Algebraic Topology, Aarhus, 1962, pp. 2–9 (see also J. F. Adams' *Algebraic Topology – A Student's Guide*, Cambridge University Press 1972, 167–177).

[38] C. H. Dowker, *Topology of metric complexes*, Amer. J. Math. **75** (1952), 555–577.

[39] A. Dranishnikov and T. Januszkiewicz, *Every Coxeter group acts amenably on a compact space*, Proceedings of the 1999 Topology and Dynamics Conference (Salt Lake City, UT); Topology Proc. **24** (1999), Spring, 135–141.

[40] M. J. Dunwoody, *Accessibility and groups of cohomological dimension one*, Proc. London Math. Soc. **38** (1979), 193–215.

[41] B. Eckmann, *Cyclic homology and the Bass conjecture,* Comment. Math. Helv. **61** (1986), 193–202.

[42] B. Eckmann, *Introduction to ℓ_2-methods in topology: reduced ℓ_2-homology, harmonic chains and ℓ_2-Betti numbers,* Israel Journal of Mathematics **117** (2000), 183–219.

[43] B. Eckmann, *Idempotents in a complex group algebra, projective modules, and the von Neumann algebra,* Arch. Math. **76** (2001), 241–249.

[44] A. D. Elmendorf, I. Kriz, M. A. Mandell and J. P. May, *Modern Foundation for Stable Homotopy Theory,* in *Handbook of Algebraic Topology (Ed. I. M. James),* Elsevier 1995, 213–253.

[45] I. Emmanouil, *On a class of groups satisfying Bass' conjecture,* Invent. Math. **132** (1998), 307–330.

[46] M. Falk and R. Randell, *Pure braid groups and products of free groups,* Contemp. Math. **78** (1988), 217–228.

[47] J. Faraut and K. Harzallah, *Distances hilbertiennes invariantes sur un espace homogène,* Ann. Inst. Fourier **24** (1974), 171–217.

[48] M. Farber and S. Weinberger, *On the zero-in-the-spectrum conjecture,* Ann. of Math. (2), **154** (2001) no. 1, 139–154.

[49] D. S. Farley, *A proper isometric action of Thompson's group V on Hilbert space,* preprint 2001.

[50] F. T. Farrell and P. A. Linnell, *Whitehead groups and the Bass conjecture*, preprint 2000.

[51] R. I. Grigorchuk, P. Linnell, T. Schick and A. Zuk, *On a conjecture of Atiyah,* C. R. Acad. Sci. Paris, **331**, Série I (2000), 663–668.

[52] A. Guichardet, *Cohomologie des groupes topologiques et des algèbres de Lie,* Cedic-F. Nathan, Paris 1980.

[53] U. Haagerup, *An example of a non-nuclear C^*-algebra which has the metric approximation property,* Invent. Math. **50** (1979), 279–293.

[54] I. Hambleton and E. K. Pedersen, *Identifying assembly maps in K- and L-theory,* preprint 2001.

[55] P. de la Harpe, *Groupes hyperboliques, algèbres d'opérateurs et un théorème de Jolissaint,* C. R. Acad. Sci. Paris, **307** (1988), 771–774.

[56] P. de la Harpe and A. Valette, *La propriété (T) de Kazhdan pour les groupes localement compacts,* Astérisque **175**, Soc. Math. France 1989.

[57] A. Hattori, *Rank element of a projective module,* Nagoya J. Math. **25** (1965), 113–120.

[58] N. Higson, *Bivariant K-theory and the Novikov conjecture,* Geom. Funct. Anal. **10** (2000), 563–581.

[59] N. Higson and G. Kasparov, *Operator K-theory for groups which act properly and isometrically on Euclidean space,* Electron. Res. Announc. Amer. Math. Soc. **3** (1997), 131–142.

[60] N. Higson, V. Lafforgue and G. Skandalis, *Counterexamples to the Baum–Connes Conjecture,* preprint 2001.

[61] N. Higson, E. K. Pedersen and J. Roe, *C^*-algebras and controlled topology,* K-theory **11** (1997), 209–239.

[62] N. Higson and J. Roe, *Amenable group actions and the Novikov conjecture,* J. reine angew. Math. **519** (2000), 143–153.

[63] N. Higson and J. Roe, *Analytic K-Homology,* Oxford Mathematical Monographs, Oxford University Press, 2000.

[64] N. Higson, J. Roe and T. Schick, *Spaces with vanishing ℓ_2-homology and their fundamental groups (after Farber and Weinberger),* Geom. Dedicata **87** (2001) no.1–3, 335–343.

[65] M. Jakob, *A bordism-type description of homology,* Manuscripta Math. **96** (1998), 67–80.

[66] P. Jolissaint, *Rapidly decreasing functions in reduced C^*-algebras of groups,* Trans. Amer. Math. Soc. **317** (1990), 167–196.

[67] P. Julg, *Travaux de N. Higson et G. Kasparov sur la conjecture de Baum–Connes,* Séminaire BOURBAKI, Astérisque **252** (1998), Exp. No. 841, 4, 151–183.

[68] D. M. Kan and W. P. Thurston, *Every connected space has the homology of a $K(\pi,1)$,* Topology **15** (1976), 253–258.

[69] M. Karoubi, *K-Theory, An Introduction,* Springer Grundlehren der Mathematischen Wissenschaften, **226**, 1978.

[70] G. G. Kasparov, *Equivariant KK-theory and the Novikov conjecture,* Invent. Math. **91** (1988), 147–201.

[71] D. Krammer, *The braid group B_4 is linear,* Invent. Math. **142** (2000), 451–486.

[72] P. H. Kropholler, *Hierarchical decompositions, generalized Tate cohomology, and groups of type FP_∞,* in: (A. Duncan, N. Gilbert and J. Howie, eds.) *Proceedings of the Edinburgh Conference on Geometric Group Theory,* 1993, Cambridge U. P. Lond. Math. Soc. Lecture Note Ser. **204** (1995), 190–216.

[73] P. H. Kropholler and G. Mislin, *On groups acting on finite dimensional spaces with finite stabilizers,* Comment. Math. Helv. **73** (1998), 122–136.

[74] V. Lafforgue, *Une démonstration de la conjecture de Baum–Connes pour les groupes réductifs sur un corps p-adique et pour certains groups discrets possédant la propriété (T),* C.R. Acad. Sci. Paris Sér. I Math. **327** (1998), no. 5, 439–444.

[75] V. Lafforgue, *A proof of property (RD) for cocompact lattices of $SL(3,\mathbb{R})$ and $SL(3,\mathbb{C})$,* J. Lie Theory **10**, 2000, no.2, 255–267.

[76] V. Lafforgue, *K-théorie bivariante pour les algèbres de Banach et conjecture de Baum–Connes,* Invent. Math. **149** (2002), 1–95.

[77] I. J. Leary, in preparation.

[78] A. Lichnerowicz, *Spineurs harmoniques,* C. R. Acad. Sci. Paris, Sér A–B **257** (1963), 7–9.

[79] P. A. Linnell, *Decomposition of augmentation ideals and relation modules,* Proc. London Math. Soc. (3) **47**, (1983), 83–127.

[80] P. A. Linnell, *Analytic version of the zero divisor conjecture,* in *Geometry and cohomology in group theory (Durham 1994),* Cambridge University Press, **252** (1998), 209–248.

[81] P. Linnell and T. Schick, *Finite group extensions and the Atiyah conjecture,* in preparation.

[82] J. Lott, *The zero-in-the-spectrum question,* L'Enseignement Mathématique **42** (1996), 341–376.

[83] W. Lück, *Transformation Groups and Algebraic K-Theory,* Lecture Notes in Mathematics **1408**, Springer 1989.

[84] W. Lück, *Hilbert modules and modules over finite von Neumann algebras and applications to L^2-invariants,* Math. Ann. **309** (1997), 247–285.

[85] W. Lück, *Dimension theory of arbitrary modules over finite von Neumann algebras and L^2-Betti numbers I: Foundations,* J. reine angew. Math. **495** (1998), 135–162.

[86] W. Lück, *Dimension theory of arbitrary modules over finite von Neumann algebras and L^2-Betti numbers II. Applications to Grothendieck groups, L^2-Euler characteristic and Burnside groups,* J. reine angew. Math. **496** (1998), 213–236.

[87] W. Lück, *The type of the classifying space for a family of subgroups,* J. Pure Appl. Algebra **149** (2000), 177–203.

[88] W. Lück, *Chern characters for proper equivariant homology theories and applications to K- and L-theory,* J. reine angew. Math. **543** (2002), 193–234.

[89] W. Lück, *The relation between the Baum–Connes Conjecture and the Trace Conjecture,* Invent. Math. **149** (2002), 123–152.

[90] W. Lück, *L^2-invariants of Regular Coverings of Compact Manifolds and CW-Complexes,* in: *Handbook of Geometric Topology,* Elsevier Sciences **2002**, 735–817.

[91] W. Lück, *L^2-Invariants: Theory and Applications to Geometry and K-Theory,* Ergebnisse der Mathematik und ihrer Grenzgebiete **44**, Springer 2002.

[92] W. Lück and D. Meintrup, *On the universal space for group actions with compact isotropy,* Proceedings of the conference "Geometry and Topology"in Aerhus, August 1998, editors: K. Grove, I. Madsen abd E. Pedersen, Contemporary Mathematics AMS **258** (2000), 293–305.

[93] W. Lück and B. Oliver, *Chern characters for equivariant K-theory of proper G-CW-complexes,* in: *Cohomological methods in homotopy theory (Bellaterra, 1998),* Prog. Math. **196**, Birkhäuser 2001, 217–247.

[94] R. C. Lyndon and P. E. Schupp: *Combinatorial Group Theory,* Ergebnisse der Mathematik und ihrer Grenzgebiete **89**, 1977.

[95] M. Matthey, *K-theories, C^*-algebras and assembly maps,* Thesis, University of Neuchâtel (Switzerland), 2000.

[96] J. P. May, *Equivariant Homotopy and Cohomology Theory,* AMS Regional Conference Series in Mathematics **91**, 1996.

[97] J. Milnor, *Construction of universal bundles, I,* Ann. of Math. **63** (1956), 272–284.

[98] I. Mineyev and G. Yu, *The Baum–Connes conjecture for hyperbolic groups,* Invent. Math. **149** (2002), 97–122.

[99] A. S. Miscenko, *Infinite-dimensional representations of discrete groups and higher signatures,* (Russian) Izv. Akad. Nauk SSSR Ser. Mat. **38** (1974), 81–106.

[100] G. Mislin, *Mapping class groups, characteristic classes and Bernoulli numbers,* CRM Proceedings and Lecture Notes, **6** (1994), 103–131.

[101] G. Mislin, *On the classifying space for proper actions,* in: *Cohomological methods in homotopy theory (Bellaterra 1998),* Progr. Math. **196**, Birkhäuser 2001, 263–269.

[102] P. S. Mostert, *Local cross sections in locally compact groups,* Proc. Amer. Math. Soc. **4** (1953), 645–649.

[103] G. Niblo and L. Reeves, *Groups acting on* CAT(0) *cube complexes,* Geom. Topol. **1** (1997), approx. 7 pp. (electronic).

[104] B. E. A. Nucinkis, *Is there an easy algebraic characterization of universal proper G-spaces?* Manuscripta Math. **102** (2000), 335–345.

[105] H. Oyono-Oyono, *La conjecture de Baum–Connes pour les groupes agissant sur les arbres,* C.R. Acad. Sci. Paris, Sér. I **326** (1998), 799–804.

[106] H. Oyono-Oyono, *Baum–Connes conjecture and extensions,* J. reine angew. Math. **532** (2001), 133–149.

[107] G. K. Pedersen, *C^*-algebras and their automorphism groups,* Academic Press, 1979.

[108] F. Peterson, *Some remarks on Chern classes*, Annals of Math. **69** (1959), 414–420.

[109] M. V. Pimsner, *KK-groups of crossed products by groups acting on trees,* Invent. Math. **86** (1986), 603–634.

[110] R. T. Powers, *Simplicity of the C^*-algebra associated with the free group on two generators,* Duke J. Math. **42** (1975), 151–156.

[111] D. Rolfsen and J. Zhu, *Braids, ordering and zero divisors,* Journal of Knot Theory and Its Ramifications, **7** No. 6 (1998), 837–841.

[112] J. Rosenberg, *C^*-algebras, positive scalar curvature, and the Novikov conjecture.* Inst. Hautes Études Sci. Publ. Math. **58** (1983), 197–212.

[113] J. Rosenberg, *Algebraic K-theory and its applications*, Graduate Text in Mathematics **147**, Springer-Verlag 1994.

[114] R. Roy, *The trace conjecture – a counterexample,* K-Theory **17** (2000), 209–213.

[115] J. A. Schafer, *Relative cyclic homology and the Bass conjecture,* Comment. Math. Helv. **67** (1992), 214–225.

[116] J. A. Schafer, *The Bass Conjecture and Group von Neumann Algebras*, K-Theory **19** (2000), 211–217.

[117] T. Schick, *A counterexample to the (unstable) Gromov–Lawson–Rosenberg conjecture,* Topology **37** (1998), 1165–1168.

[118] T. Schick, *Finite group extensions and the Baum–Connes conjecture,* preprint 2000.

[119] T. Schick, *The strong Bass conjecture for group elements of finite order and for residually finite groups,* preprint 2000.

[120] T. Schick, *Integrality of L^2-Betti numbers,* Math. Ann. **317** (2000), 727–750.

[121] J.-P. Serre, *Linear representations of finite groups*, Springer-Verlag 1977.

[122] J.-P. Serre, *Trees*, Springer-Verlag 1980.

[123] J.-P. Serre, *Galois Cohomology,* Springer-Verlag 1997.

[124] J. Slominska, *On the equivariant Chern homomorphisms*, Bull. Acad. Pol. **24** (1976), 909–913.

[125] E. H. Spanier, *Algebraic Topology*, McGraw-Hill 1966.

[126] J. Stallings, *Centerless groups – an algebraic formulation of Gottlieb's theorem,* Topology, **4** (1965), 129–134.

[127] S. Stolz, *Survey on the problem of finding a positive scalar curvature metric on a closed manifold,* in preparation.

[128] R. M. Switzer, *Algebraic Topology – Homotopy and Homology,* Grundlehren **212**, Springer Verlag 1973.

[129] J. L. Tu, *The Baum–Connes conjecture and discrete group actions on trees,* K-Theory **17** no.4 (1999), 303–318.

[130] A. Valette, *Introduction to the Baum–Connes Conjecture*, Lecture Notes in Mathematics, ETH Zürich, Birkhäuser Verlag 2002.

[131] R. Wood, *Banach algebras and Bott periodicity,* Topology **4** (1966), 371–389.

[132] Z. Yoshimura, *A note on complex K-theory of infinite CW-complexes*, Journal of the Mathematical Society of Japan, **26**, No.2 (1974), 289–295.

On the Baum–Connes Assembly Map for Discrete Groups

Alain Valette
With an Appendix by Dan Kucerovsky

Abstract: In these notes, we study the Baum–Connes analytical assembly maps (or index maps) $\mu_i^\Gamma : RK_i^\Gamma(\underline{E}\Gamma) \to K_i(C_r^*\Gamma)$ and $\tilde{\mu}_i^\Gamma : RK_i^\Gamma(\underline{E}\Gamma) \to K_i(C^*\Gamma)$, for a countable group Γ. Here $RK_i^\Gamma(\underline{E}\Gamma)$ denotes the Γ-equivariant K-homology with Γ-compact supports of the universal space $\underline{E}\Gamma$ for proper Γ-actions, while $K_i(C_r^*\Gamma)$ (resp. $K_i(C^*\Gamma)$) denotes the analytical K-theory of the reduced (resp. full) C*-algebra of Γ. As it is simple and direct, we use the definition of μ_i^Γ suggested by Baum, Connes and Higson in Section 3 of [BCH94]. The Baum–Connes conjecture asserts that, for any group Γ, the map μ_i^Γ is an isomorphism ($i = 0, 1$). The contents of this paper are as follows:

1. We make the necessary changes for constructing $\tilde{\mu}_i^\Gamma$, and give a detailed proof that μ_i^Γ and $\tilde{\mu}_i^\Gamma$ provide K-theory elements of the corresponding C*-algebras.
2. We carefully describe the behaviour of the left-hand side of the assembly maps under group homomorphisms, and we prove that $\tilde{\mu}_i^\Gamma$ is natural with respect to arbitrary group homomorphisms. As a consequence, we get a new proof of the fact that, if Γ acts freely on the space X, then the equivariant K-homology $K_*^\Gamma(X)$ is isomorphic to the K-homology $K_*(\Gamma\backslash X)$ of the orbit space.
3. To illustrate the non-triviality of the assembly map, we give a direct proof of the Baum–Connes conjecture for the group $\mathbb{Z}$ of integers, not appealing to equivariant KK-theory.
4. Denote by $\tilde{\kappa}_\Gamma : \Gamma \to K_1(C_r^*\Gamma)$ the homomorphism induced by the canonical inclusion of Γ in the unitary group of $C_r^*\Gamma$. We show that there exists a homomorphism $\tilde{\beta}_t : \Gamma \to RK_1^\Gamma(\underline{E}\Gamma)$ such that $\tilde{\kappa}_\Gamma = \mu_i^\Gamma \circ \tilde{\beta}_t$; this extends a result of Natsume [Nat88] for Γ torsion-free.

The Appendix, by Dan Kucerovsky, discusses the assembly map in terms of unbounded K-homology elements.

1. Introduction

1.1. The Baum–Connes conjecture

Let Γ be a countable, discrete group. The *Baum–Connes conjecture* is a tantalizing programme that identifies two objects associated with Γ, one analytical and one geometrical or topological.

The analytical side involves the K-theory of the *reduced C*-algebra* $C_r^*\Gamma$, which is the C*-algebra generated by Γ in its left regular representation on the Hilbert space $\ell^2(\Gamma)$. The K-theory used here, $K_i(C_r^*\Gamma)$ for $i = 0, 1$, is the usual K-theory for Banach algebras, as described e.g. in [Tay75].

On the opposite side, one finds the K-homology (with compact supports) of a certain classifying space. More precisely, consider the universal space $\underline{E}\Gamma$ for proper Γ-actions (as described in (1.6) of [BCH94], see also Section 2 below; such a space is unique up to Γ-equivariant homotopy). A Γ-invariant subset $Y \subset \underline{E}\Gamma$ is *Γ-compact* if the orbit space $\Gamma\backslash Y$ is compact. The geometric group is the Γ-equivariant K-homology with Γ-compact supports $RK_i^\Gamma(\underline{E}\Gamma)$ of $\underline{E}\Gamma$, i.e. the inductive limit of the Γ-equivariant K-homology [1] groups $K_i^\Gamma(Y)$, where Y runs along Γ-compact subsets of $\underline{E}\Gamma$.

The link between both sides of the conjecture is provided by the *analytic assembly map*, or *index map*

$$\mu_i^\Gamma : RK_i^\Gamma(\underline{E}\Gamma) \to K_i(C_r^*\Gamma)$$

($i = 0, 1$). The definition of the assembly map can be traced back to a result of Kasparov [Kas83]: suppose that Z is a proper Γ-compact manifold endowed with a Γ-invariant elliptic differential operator D on some Γ-vector bundle over Z. Then, in spite of the non-compactness of the manifold Z, the *index* of D has a well-defined meaning as an element of the K-theory $K_i(C_r^*\Gamma)$. On the other hand, using the universal property of $\underline{E}\Gamma$, the manifold Z maps continuously Γ-equivariantly to $\underline{E}\Gamma$, and the pair (Z, D) defines an element of the equivariant K-homology with compact supports $RK_i^\Gamma(\underline{E}\Gamma)$. Then, one sets

$$\mu_i^\Gamma(Z, D) = Index(D).$$

Elaborating on this, and using the concept of abstract elliptic operator (or Kasparov triple), one defines the assembly map μ_i^Γ, which is a group homomorphism.

Conjecture 1. *(the Baum–Connes conjecture) For $i = 0, 1$, the assembly map*

$$\mu_i^\Gamma : RK_i^\Gamma(\underline{E}\Gamma) \to K_i(C_r^*\Gamma)$$

is an isomorphism.

[1] The groups $K_i^\Gamma(Y)$ can be defined as the equivariant Kasparov groups $KK_i^\Gamma(C_0(Y), \mathbb{C})$, where $C_0(Y)$ denotes the abelian C*-algebra of continuous functions vanishing at infinity on Y; for equivariant Kasparov theory, we refer to [Kas95], [Kas88]. We shall give more details on that definition in Section 2.2. As in Chapter 5 of [Roe96], it is also possible to define $K_i^\Gamma(Y)$ as the *K-theory* of the algebra of pseudo-local operators modulo locally compact operators on Y, in a suitable covariant representation of $C_0(Y)$.

This conjecture is part of a more general conjecture (discussed in [BCH94]) where discrete groups are replaced by arbitrary locally compact groups [2]. The reason for restricting to discrete groups is that, in a sense, this case is both interesting and difficult. The main difficulty comes from the analytical side: e.g., there is no general structure result for the reduced C*-algebra of a discrete group, so that its K-theory is usually quite hard to compute (recall that, in many important cases, e.g. lattices in semi-simple Lie group, $C_r^*\Gamma$ is actually *simple*, see [BCdlH94]). The interest of Conjecture 1 is that it *implies* several other famous conjectures in topology, geometry, algebra and functional analysis.

Conjecture 2. *(the Novikov conjecture) For closed oriented manifolds with fundamental group* Γ*, the higher signatures coming from* $H^*(\Gamma, \mathbb{Q})$ *are oriented homotopy invariants.*

The Novikov conjecture follows from the rational injectivity of μ_i^Γ (see [BCH94], Theorem 7.11; [FRR95], Section 6).

Conjecture 3. *(one direction of the Gromov–Lawson–Rosenberg conjecture) If* M *is a closed spin manifold with fundamental group* Γ*, and if* M *is endowed with a metric of positive scalar curvature, then all higher* $\hat{A}$*-genera (coming from* $H^*(\Gamma, \mathbb{Q})$*) do vanish.*

Conjecture 3 is also a consequence of the rational injectivity of μ_i^Γ (see Theorem 7.11 in [BCH94]).

Let us also mention the conjecture of idempotents for $C_r^*\Gamma$; since $C_r^*\Gamma$ is a completion of the complex group algebra $\mathbb{C}\Gamma$, this conjecture is stronger than the classical conjecture of idempotents, discussed e.g. in [Pas85].

Conjecture 4. *(the conjecture of idempotents, or Kaplansky–Kadison conjecture) Let* Γ *be a torsion-free group. Then* $C_r^*\Gamma$ *has no idempotent other than 0 or 1.*

This conjecture would follow from the surjectivity of μ_0^Γ (see Proposition 7.16 in [BCH94]; Proposition 3 in [Val89]).

It has to be emphasized that Conjecture 1 makes $K_i(C_r^*\Gamma)$ computable, at least up to torsion. The reason is that $RK_i^\Gamma(\underline{E}\Gamma)$ is computable up to torsion. Let us explain this briefly. Let $F\Gamma$ be the space of complex-valued functions on Γ, whose support is finite and contained in the set of torsion elements of Γ. Letting Γ act by conjugation on torsion elements, $F\Gamma$ becomes a Γ-module; denote by $H_j(\Gamma, F\Gamma)$ the j-th homology group of Γ with coefficients in $F\Gamma$. In Section 6 of [BC88a], Baum and Connes define a *Chern character*

$$ch_\Gamma : RK_i^\Gamma(\underline{E}\Gamma) \to \bigoplus_{n=0}^{\infty} H_{i+2n}(\Gamma, F\Gamma),$$

[2]Not to mention an even more general conjecture, that we deliberately ignore here, concerning either locally compact groups acting on locally compact spaces or foliated manifolds, and with coefficients in an arbitrary auxiliary C*-algebra.

and state in Proposition 15.2 of [BC88a] that the Chern character is an isomorphism after tensoring by $\mathbb{C}$, i.e.

$$ch_\Gamma \otimes 1 : RK_i^\Gamma(\underline{E}\Gamma) \otimes_{\mathbb{Z}} \mathbb{C} \to \bigoplus_{n=0}^{\infty} H_{i+2n}(\Gamma, F\Gamma)$$

is an isomorphism (another Chern character, having all the desired properties, has been constructed by Matthey in Theorem 1.4 of [Mat]; he conjectures that his Chern character coincides with Baum–Connes', and proves this for $\Gamma = G \times \mathbb{Z}/n\mathbb{Z}$, with BG a closed manifold).

As an example, consider the case where Γ is torsion-free. The Γ-module $F\Gamma$ is just the trivial module $\mathbb{C}$; on the other hand, let $B\Gamma$ be a classifying space for Γ, i.e. a $K(\Gamma, 1)$-space; let $E\Gamma$ be its universal covering space. Since Γ is torsion-free, any proper action is automatically free, so we may take $\underline{E}\Gamma = E\Gamma$. Then there is a canonical isomorphism

$$RK_i^\Gamma(E\Gamma) \simeq RK_i(B\Gamma),$$

where $RK_i(B\Gamma)$ denotes the K-homology with compact supports of $B\Gamma$. This identification is compatible with the usual Chern character in K-homology, i.e. there is a commutative diagram

$$\begin{array}{ccc} RK_i^\Gamma(E\Gamma) & \xrightarrow{\simeq} & RK_i(B\Gamma) \\ & {}_{ch_\Gamma}\searrow \quad \swarrow_{ch} & \\ & \bigoplus_{n=0}^{\infty} H_{i+2n}(\Gamma, \mathbb{C}) & \end{array}$$

(see [BCH94], p. 274; [Mat], Theorem 1.4).

As another example, take for Γ a finite group. Then

$$RK_0^\Gamma(\underline{E}\Gamma) = K_0^\Gamma(pt) \simeq R(\Gamma),$$

where $R(\Gamma)$ is the representation ring of Γ. On the other hand,

$$\bigoplus_{n=0}^{\infty} H_{2n}(\Gamma, F\Gamma) = H_0(\Gamma, F\Gamma)$$

is the complex vector space on the set of conjugacy classes of Γ. In other words, the fact that the Chern character is an isomorphism (after tensoring with $\mathbb{C}$) incorporates the classical but not quite obvious fact that, for a finite group, the number of irreducible representations is equal to the number of conjugacy classes.

1.2. What these Notes are about

The basic sources for the Baum–Connes conjecture are the original articles [BC00], [BC88a], [BC88b], [BCH94]; in textbooks, see various sections in the books by Connes [Con94] and by Higson–Roe [HR00b]. For expository presentations entirely devoted to the Baum–Connes conjecture, see the Bourbaki seminars by Julg [Jul98] and Skandalis [Ska99], and the book by the author [Val02]. In particular, the last three references contain relevant information about the status of the Baum–Connes conjecture, especially for which classes of discrete groups surjectivity and/or injectivity of the assembly map has been proved.

These Notes were begun in 1996–97, and have gone through many states since then. At the origin, they were aimed at backing up some aspects of the proof of the Baum–Connes conjecture for one-relator groups in [BBV99]. Publication was delayed, due to the feeling that the results contained in these Notes were known to every expert. Part of the Notes was used as we were working on [Val02], which explains some overlap, for which we apologize. Eventually I yielded to the friendly insistance of some colleagues and some graduate students, who convinced me that these Notes, although not fully original, could be of some interest for beginners. Let us discuss now the content of these Notes.

It has been noticed by many authors (see e.g. [Con94] p. 99, [FRR95] Section 6, [Jul98]) that the analytical assembly map $\mu_i^\Gamma : RK_i^\Gamma(\underline{E}\Gamma) \to K_i(C_r^*\Gamma)$ factors through the K-theory of the *full* C*-algebra $C^*\Gamma$, which is the universal C*-completion of the group algebra $\mathbb{C}\Gamma$. More precisely, there is a homomorphism

$$\tilde{\mu}_i^\Gamma : RK_i^\Gamma(\underline{E}\Gamma) \to K_i(C^*\Gamma)$$

such that

$$\mu_i^\Gamma = (\lambda_\Gamma)_* \circ \tilde{\mu}_i^\Gamma,$$

where $\lambda_\Gamma : C^*\Gamma \to C_r^*\Gamma$ is the canonical epimorphism corresponding to the left regular representation of Γ. As we shall see, the assembly map $\tilde{\mu}_i^\Gamma$ enjoys better naturality properties than μ_i^Γ.

In Section 2, we give the definition of Baum–Connes–Higson [BCH94] for μ_i^Γ and provide the necessary changes for $\tilde{\mu}_i^\Gamma$; these definitions have the advantage of being direct and avoiding the use of Kasparov products. However, it is not completely apparent from Definition (3.8) in [BCH94] that the map μ_i^Γ is well-defined and actually provides K-theory elements of $C_r^*\Gamma$. There is a number of checks to be made, for which we give the relevant details; in particular, we pay due attention to positivity questions, often overlooked. In the process, we also show that the two definitions of μ_i^Γ, given in Sections 3 and 8 of [BCH94], are truly the same: namely, for X a Γ-compact subset of $\underline{E}\Gamma$ and $x \in KK_i^\Gamma(C_0(X), \mathbb{C})$, then $\mu_i^\Gamma(x)$ can be defined by first applying Kasparov's descent homomorphism j_Γ to x, and then taking the Kasparov product of $j_\Gamma(x)$ by the canonical line bundle $[\mathcal{L}_X]$ over $C_0(X) \rtimes \Gamma$.

Next we will prove that the analytic assembly map $\tilde{\mu}_i^\Gamma$ is natural, i.e.:

Theorem 1.1. *Let $\alpha : \Gamma_1 \to \Gamma_2$ be a group homomorphism; then there is a commutative diagram:*

$$\begin{array}{ccc} RK_i^{\Gamma_1}(\underline{E}\Gamma_1) & \xrightarrow{\tilde{\mu}_i^{\Gamma_1}} & K_i(C^*\Gamma_1) \\ \downarrow{\alpha_*} & & \downarrow{\alpha_*} \\ RK_i^{\Gamma_2}(\underline{E}\Gamma_2) & \xrightarrow{\tilde{\mu}_i^{\Gamma_2}} & K_i(C^*\Gamma_2) \end{array}$$

This result is of course known to experts (see [Con94], pp. 96–97; [FRR95], p. 44; [Ros83]; [LÖ2]); we thought it worthwhile to record a proof. Most of Section 3 is devoted to the proof of Theorem 1.1: 3.1 deals with the case of group monomorphisms, 3.2 with group epimorphisms (since any group homomorphism is the product of an epimorphism with a monomorphism, this clearly implies the general case). Actually one difficulty in the proof is to carefully describe the functoriality of the left-hand side, i.e. how $RK_i^\Gamma(\underline{E}\Gamma)$ behaves under group homomorphisms. For an epimorphism $\alpha : \Gamma_1 \twoheadrightarrow \Gamma_2$ with kernel N, the construction is inspired from Kasparov's descent homomorphism (Theorem 3.4 in [Kas88]): if Γ_1 acts properly on a space X, this provides a homomorphism $\alpha_* : K_i^{\Gamma_1}(X) \to K_i^{\Gamma_2}(N\backslash X)$. In particular, if Γ acts *freely* on X, by considering the constant homomorphism α from Γ to the trivial group, we get an explicit map $\alpha_* : K_i^\Gamma(X) \to K_i(\Gamma\backslash X)$ which, as shown in Corollary 3.7, coincides with the isomorphism constructed classically using Morita equivalence (see [Gre77]).

The proof of Theorem 1.1 uses the fact that the full group C*-algebra is functorial for arbitrary group homomorphisms. By way of contrast, the reduced C*-algebra is functorial only for group monomorphisms. This purely analytical reason is responsible for the limited naturality of μ_i^Γ, that we now state precisely.

Let $\alpha : \Gamma_1 \to \Gamma_2$ be a group monomorphism; then there is a commutative diagram:

$$\begin{array}{ccc} C^*\Gamma_1 & \xrightarrow{\alpha_*} & C^*\Gamma_2 \\ \lambda_{\Gamma_1}\downarrow & & \downarrow\lambda_{\Gamma_2} \\ C_r^*\Gamma_1 & \xrightarrow{\alpha_*} & C_r^*\Gamma_2 \end{array}$$

As an immediate consequence, we have:

Corollary 1.2. *For a group monomorphism $\alpha : \Gamma_1 \to \Gamma_2$, there is a commutative diagram:*

$$\begin{array}{ccc} RK_i^{\Gamma_1}(\underline{E}\Gamma_1) & \xrightarrow{\mu_i^{\Gamma_1}} & K_i(C_r^*\Gamma_1) \\ \alpha_*\downarrow & & \downarrow\alpha_* \\ RK_i^{\Gamma_2}(\underline{E}\Gamma_2) & \xrightarrow{\mu_i^{\Gamma_2}} & K_i(C_r^*\Gamma_2) \end{array}$$

It is however possible to restore full naturality by imposing conditions on the source group. Recall that a group Γ is *amenable* if $\lambda_\Gamma : C^*\Gamma \to C_r^*\Gamma$ is an isomorphism (this is one among lots of equivalent definitions, see [Ped79] 7.3). More generally, Γ is *K-amenable* in the sense of Cuntz [Cun83] if

$$\lambda_\Gamma^* : K^0(C_r^*\Gamma) \to K^0(C^*\Gamma)$$

is an isomorphism in K-homology; this is known to imply that the K-theory map

$$(\lambda_\Gamma)_* : K_i(C^*\Gamma) \to K_i(C_r^*\Gamma)$$

is an isomorphism (see [Cun83], Theorem 2.1). Until Lafforgue's remarkable results [Laf98], all the groups for which Conjecture 1 was proved, belonged to the class of K-amenable groups.

Corollary 1.3. *Let $\alpha : \Gamma_1 \to \Gamma_2$ be a group homomorphism, where Γ_1 is K-amenable; then there is a commutative diagram:*

$$\begin{array}{ccc} RK_i^{\Gamma_1}(\underline{E}\Gamma_1) & \xrightarrow{\mu_i^{\Gamma_1}} & K_i(C_r^*\Gamma_1) \\ \alpha_* \downarrow & & \downarrow \alpha_* \\ RK_i^{\Gamma_2}(\underline{E}\Gamma_2) & \xrightarrow{\mu_i^{\Gamma_2}} & K_i(C_r^*\Gamma_2) \end{array}$$

At this juncture, notice that, since the left-hand side of the Baum–Connes conjecture (i.e. the geometric group $RK_i^\Gamma(\underline{E}\Gamma)$) is fully functorial with respect to group homomorphisms, it would follow from the *truth* of the Baum–Connes conjecture that the right-hand side is fully functorial, i.e. for any group homomorphism $\alpha : \Gamma_1 \to \Gamma_2$ there should exist a functorial $\alpha_* : K_i(C_r^*\Gamma_1) \to K_i(C_r^*\Gamma_2)$; this functoriality, on which we elaborate in Example 3.6, can be "explained" by a conjecture of J.-B. Bost (see [Ska99]): the range of the Baum–Connes assembly map μ_*^Γ should be the K-theory of $\ell^1\Gamma$ rather than the one of $C_r^*\Gamma$; and of course $\ell^1\Gamma$ is fully functorial with respect to group homomorphisms.

To illustrate the construction of the analytical assembly map, we give in Section 4 a direct proof of the fact that

$$\mu_1^{\mathbb{Z}} : RK_1^{\mathbb{Z}}(\underline{E}\mathbb{Z}) \to K_1(C^*\mathbb{Z})$$

is an isomorphism. Of course, it is well-known that the group $\mathbb{Z}$ satisfies the Baum–Connes conjecture, and it might even be tempting to believe that this result is obvious (this is essentially the opinion expressed in Lemma 3.5 of [BC88b]). I think that, although both groups involved are isomorphic to $\mathbb{Z}$, the definition of the assembly map is intricate enough, so that one really has to check that $\mu_1^{\mathbb{Z}}$ maps generator to generator. It is informative here to look at Kasparov's dual map

$$\alpha : K^i(C^*\Gamma) \to RK^i(B\Gamma),$$

defined in [Kas75], Section 8; [Kas95], Section 9; [Kas88], Section 6, and which was considered prior to the Baum–Connes assembly map. Then

$$\alpha : K^1(C^*\mathbb{Z}) \to K^1(B\mathbb{Z})$$

is an isomorphism: the non-triviality of this fact is apparent from the proofs in [Kas75], Theorem 1 of Section 8; [Ros84], Lemma 3.2. Coming back to "$\mu_1^{\mathbb{Z}}$ is an isomorphism", it is clear that this result is contained in Kasparov's conspectus [Kas95], but hidden in the wide generality of Theorem 1 of Section 7. What is explicit there (and non-trivial) is the fact the Connes–Kasparov conjecture holds for the 1-dimensional Lie group $\mathbb{R}$ ([Kas95], Lemma 4 in Section 5). Then one appeals to the machinery of equivariant KK-theory, whose powerful functorialities allow to descend from a Lie group to a discrete subgroup. I thought that it was worthwhile to give a direct proof. Another direct proof can be found in Example 12.5.9 of the recent book [HR00b].

In the final Section 5, we consider the canonical homomorphism

$$\tilde{\kappa}_\Gamma : \Gamma \to K_1(C_r^*\Gamma)$$

obtained from the canonical embedding of Γ into the unitary group of $C_r^*\Gamma$. Since $K_1(C_r^*\Gamma)$ is abelian, $\tilde{\kappa}_\Gamma$ factors through a homomorphism

$$\kappa_\Gamma : \Gamma^{ab} \to K_1(C_r^*\Gamma)$$

where Γ^{ab} denotes the abelianization of Γ. It is known (see [EN87], [BV96]) that κ_Γ is always rationally injective.

Theorem 1.4. *There exists a homomorphism*

$$\beta_t : \Gamma^{ab} \to RK_1^\Gamma(\underline{E}\Gamma)$$

such that, as homomorphisms $\Gamma^{ab} \to K_1(C_r^*\Gamma)$, *one has*

$$\kappa_\Gamma = \mu_1^\Gamma \circ \beta_t.$$

For Γ torsion-free, such a map β_t was previously constructed by Natsume [Nat88] who, however, does not give the proof that β_t is a group homomorphism. The proof of Theorem 1.4 appeals to Theorem 1.1 together with the fact (proved in Section 4) that conjecture 1 holds for the group of integers. Recall that $\Gamma^{ab} = H_1(\Gamma, \mathbb{Z})$, and that the inclusion $\mathbb{C} \hookrightarrow F\Gamma$ associated with the trivial conjugacy class in Γ, induces an inclusion

$$\Gamma^{ab} \otimes_{\mathbb{Z}} \mathbb{C} = H_1(\Gamma, \mathbb{C}) \hookrightarrow \bigoplus_{n=0}^{\infty} H_{2n+1}(\Gamma, F\Gamma) \simeq RK_1^\Gamma(\underline{E}\Gamma) \otimes_{\mathbb{Z}} \mathbb{C}.$$

Theorem 1.4 together with the rational injectivity of κ_Γ then imply that μ_1^Γ is rationally injective on the image of β_t, i.e. on the lowest dimensional part of $RK_1^\Gamma(\underline{E}\Gamma)$.

Deep generalizations of Theorem 1.4 have been proposed in Matthey's PhD thesis [Mat00], in the form of a "delocalized" version allowing him to treat other conjugacy classes in Γ than the trivial one. More precisely, for $0 \leq j \leq 2$, he constructs maps $\beta_t : H_j(\Gamma, F\Gamma) \to RK_j^\Gamma(\underline{E}\Gamma) \otimes_{\mathbb{Z}} \mathbb{C}$ and $\beta_a : H_j(\Gamma, F\Gamma) \to K_j(C_r^*\Gamma) \otimes_{\mathbb{Z}} \mathbb{C}$, commuting with $\mu_j^\Gamma \otimes 1$.

The Appendix, by Dan Kucerovsky, presents the construction of the Baum–Connes assembly map "in the unbounded picture", i.e. when the K-homology elements in $RK_*^\Gamma(\underline{E}\Gamma)$ are given by unbounded Kasparov elements (as in [BJ83], [Kuc94]). The interest of this approach is that most K-homology elements of geometric origin are given by unbounded operators: e.g. first order, elliptic, differential operators (like de Rham, Dirac, or signature operators) define unbounded operators on the Hilbert space of L^2-sections of the corresponding vector bundles over the underlying manifold.

1.3. Other descriptions of assembly maps

Other approaches to the Baum–Connes assembly map have been proposed.

- For an arbitrary metric space X, J. Roe and N. Higson (see [HR00a], [Roe96]) define the coarse K-homology of X, denoted by $KX_*(X)$, the C*-algebra C^*X of operators with finite propagation on X, and the coarse assembly map
$$A_\infty : KX_*(X) \to K_*(C^*X).$$
Note that $KX_*(X) = RK_*(X)$ for X uniformly contractible with bounded geometry ([HR00a], Proposition 3.8). The *coarse Baum–Connes conjecture* is the statement that, for X a complete path metric space with bounded geometry, the map A_∞ is an isomorphism. For Γ a finitely generated group, view a Cayley graph $|\Gamma|$ as a complete path metric space. At least when Γ admits a finite complex as a classifying space, there is a "descent principle" allowing to deduce, from the conjectured isomorphism $A_\infty : KX_*(|\Gamma|) \to K_*(C^*|\Gamma|)$, the injectivity of the Baum–Connes assembly map $\mu_*^\Gamma : RK_*(B\Gamma) \to K_*(C_r^*\Gamma)$ (see [Roe96], Theorem 8.4). A comparison between this approach and the "classical" one can be found in [Roe].[3]

- J.F. Davis and W. Lück [DL98] give a categorical definition of assembly maps in algebraic K-theory, topological K-theory and L-theory, by means of spectra over the orbit category of the group Γ. The source of the Baum–Connes assembly map is defined there by considering the "orbit category" $Or(\Gamma, \mathcal{F}in)$ of quotients of Γ by finite subgroups, applying the functor $K^{top} : Or(\Gamma, \mathcal{F}in) \to SPECTRA$ constructed in Section 2 of [DL98], considering the classifying space $E(\Gamma, \mathcal{F}in)$ of $Or(\Gamma, \mathcal{F}in)$, forming the "tensor product" spectrum $E(\Gamma, \mathcal{F}in) \otimes_{Or(\Gamma,\mathcal{F}in)} K^{top}$, as in Section 1 of [DL98], and finally applying homotopy groups. W. Lück has communicated to us a simple proof [Luc] of the naturality of the source of the Baum–Connes assembly map under arbitrary group homomorphisms: when expressed in that language, it basically boils down to the fact that the orbit category is natural, i.e. that group homomorphisms map finite subgroups to finite subgroups! In Section 5 of [DL98], Davis and Lück construct an "assembly map" to $K_*(C_r^*\Gamma)$: it was recently proved by I. Hambleton and E.K. Pedersen (see [HP], Corollary 7.4), that this construction is equivalent to the one in [BCH94].

---

[3] Counterexamples to the coarse Baum–Connes conjecture, as well as candidates for counterexamples to the Baum–Connes conjecture with coefficients, have been constructed by Higson, Lafforgue and Skandalis [HLS02].

1.4. Remarks on the background

The reader of this paper is advised to have some background in C*-algebras. The reason is not that the author has been educated in the C*-faith, but rather that the difficulties encountered are analytical in nature – even to describe the functoriality of the left-hand side of the Baum–Connes conjecture! I shall use freely those parts of C*-algebra theory relevant for Kasparov's KK-theory, namely positivity, representations, and multipliers; they can be found either in Arveson's book [Arv76] or in Pedersen's book [Ped79]; for group C*-algebras, Dixmier's book [Dix77] is compulsory; for Hilbert C*-modules, I recommand Lance's lovely little book [Lan95]; for KK-theory itself, I suggest the book [JT91] by Knudsen Jensen and Thomsen.

1.5. Acknowledgements

I thank M.E.B. Bekka, P. Julg, W. Lück and G. Mislin for a number of useful conversations, and S. Echterhoff for his help in the proof of Corollary 3.7. Also, my warmest thanks to my former PhD students: C. Béguin and H. Bettaieb for digesting the first draft; I. Chatterji and especially M. Matthey for numerous useful comments on the final versions.

D. Kucerovsky kindly allowed me to append to these Notes his discussion of the unbounded picture for the assembly map.

Finally, it is a pleasure to thank the CRM at Barcelona for offering me to publish this text in this *Advanced Course in Mathematics* originating from the Euro Summer School on proper group actions, held in September 2001. Although held in the wake of September 11, 2001, that summer school remains a pleasant memory for me, for the lively atmosphere and excellent working conditions: let all the participants, and staff at CRM, be sincerely thanked here.

2. The Analytical Assembly Map

2.1. Proper actions

Let X be a metrizable space on which the group Γ acts by homeomorphisms.

Definition 2.1. The Γ-space X is *proper* if the quotient space $\Gamma\backslash X$ is Hausdorff and every point in X admits a Γ-invariant open neighbourhood which maps Γ-equivariantly continuously to an homogeneous space Γ/H, where H is a finite subgroup of Γ.

This definition is *stronger* than the usual definition of a proper action, which requires that, for any compact subsets K, L of X, the set

$$\{\gamma \in \Gamma : \gamma K \cap L \neq \emptyset\}$$

is finite (or, equivalently, the action map $\Gamma \times X \to X \times X : (\gamma, x) \mapsto (\gamma.x, x)$ is proper in the usual sense that the inverse image of a compact subset is compact). For locally compact spaces, Definition 2.1 is actually equivalent to the classical one (see [Pal61]).

According to our definition, a proper Γ-space is locally of the form $\Gamma \times_H Y$, a space induced from the action of a finite subgroup H on a space Y. We say that a proper Γ-space X is Γ-*compact* if $\Gamma \backslash X$ is compact; note that a proper, Γ-compact space has to be locally compact.

Definition 2.2. A proper Γ-space $\underline{E}\Gamma$ is *universal* if it satisfies the following universal property: for every proper Γ-space X, there exists a Γ-equivariant continuous map $X \to \underline{E}\Gamma$, and any two such maps are Γ-equivariantly homotopic.

The Γ-equivariant map $X \to \underline{E}\Gamma$ is not proper in general, but it is proper as soon as X is Γ-compact (see Lemma 3.2 below). It is clear from Definition 2.2 that a universal proper Γ-space is unique, up to Γ-equivariant homotopy. If Γ is torsion-free, any proper Γ-action is free, so we may take for $\underline{E}\Gamma$ the universal covering space $E\Gamma$ of the classifying space $B\Gamma$. If Γ is finite, we may take $\underline{E}\Gamma = pt$, the one-point space. The non-classical definition of properness in Definition 2.1 is required because there are natural examples where $\underline{E}\Gamma$ is definitely not a locally compact space. For example, the following fairly simple description of $\underline{E}\Gamma$ (see Section 2 in [BCH94]), valid for an arbitrary Γ, is not locally compact as soon as Γ is infinite. The lemma below appears in Section 2 of [BCH94]; our proof is slightly more direct than the original one.

Lemma 2.3. *Let $\underline{E}\Gamma$ be the space of finitely supported probability measures on Γ, endowed with the metric*

$$\|\mu - \nu\|_\infty = \sup\{|\mu(\gamma) - \nu(\gamma)| : \gamma \in \Gamma\},$$

and with the action of Γ given by left multiplication. Then $\underline{E}\Gamma$ is a universal proper Γ-space.

Proof. We first check that the action of Γ on $\underline{E}\Gamma$ is proper. Fix $\mu \in \underline{E}\Gamma$, and let Γ_μ be its stabilizer, a finite subgroup in Γ. Set

$$R = \inf\{\|\mu - \gamma(\mu)\|_\infty : \gamma \in \Gamma - \Gamma_\mu\};$$

one sees easily that $R > 0$. For $\epsilon > 0$, define

$$U = \{\nu \in \underline{E}\Gamma : \exists \gamma \in \Gamma : \|\nu - \gamma(\mu)\|_\infty < \epsilon\};$$

it is an open, Γ-invariant subset of $\underline{E}\Gamma$. Moreover, for $\epsilon < \frac{R}{2}$, the open set U is such that, for $\nu \in U$, the element $\gamma \in \Gamma$ with $\|\nu - \gamma(\mu)\|_\infty < \epsilon$ is unique modulo Γ_μ. Sending ν to the coset $\gamma\Gamma_\mu$ then defines a Γ-equivariant map $U \to \Gamma/\Gamma_\mu$. So $\underline{E}\Gamma$ is a proper Γ-space.

Let X be a proper Γ-space. We have to show that there exists a continuous Γ-equivariant map, which is unique up to Γ-equivariant homotopy. Uniqueness is clear, since $\underline{E}\Gamma$ is a convex set on which Γ acts affinely. For the existence, denote by W the disjoint union of the Γ/H's, where H runs along finite subgroups of Γ. Define a Γ-equivariant map $\phi : W \to \underline{E}\Gamma$ by sending the coset γH to the uniform probability measure on γH. By the lemma in Appendix 1 of [BCH94], there exists a countable partition of unity $(\alpha_k)_{k\geq 1}$ on X, consisting of Γ-invariant functions and such that, for every $k \geq 1$, there is a Γ-equivariant continuous map $\psi_k : \alpha_k^{-1}]0,1] \to W$. Then the map

$$\Psi : X \to \underline{E}\Gamma : x \mapsto \sum_{k=1}^{\infty} \alpha_k(x)(\phi \circ \psi_k)(x)$$

is continuous and Γ-equivariant. □

2.2. Equivariant K-homology

We now recall, following [BCH94], the definition of the geometric group $RK_i^\Gamma(\underline{E}\Gamma)$ that appears in the left hand side of the Baum–Connes conjecture.

Definition 2.4. The *Baum–Connes geometric group* is

$$RK_i^\Gamma(\underline{E}\Gamma) = \varinjlim_{\substack{X \subset \underline{E}\Gamma \\ X \text{ is } \Gamma-\text{compact}}} K_i^\Gamma(X),$$

where $K_i^\Gamma(X)$ is the Γ-equivariant K-homology of X.

More generally, for Y a proper Γ-space, we define $RK_i^\Gamma(Y)$ as the inductive limit of the $K_i^\Gamma(X)$'s, where X runs along Γ-compact subsets of Y. If Γ is the trivial group, we drop the superscript Γ.

An element of $K_i^\Gamma(X)$ is represented by a *Kasparov triple* or *abstract elliptic operator* $(\mathcal{H}, \pi, F)$, where:

- $\mathcal{H}$ is a Hilbert space endowed with a unitary representation of Γ;
- π is a covariant representation of $C_0(X)$ on $\mathcal{H}$, i.e. a $*$-homomorphism from $C_0(X)$ to the algebra $\mathcal{L}(H)$ of bounded operators on $\mathcal{H}$, such that

$$\pi(\gamma(f)) = \gamma\pi(f)\gamma^{-1}$$

for every $\gamma \in \Gamma$, $f \in C_0(X)$.
- F is a bounded self-adjoint operator on $\mathcal{H}$ such that $[F, \pi(f)]$, $\pi(f)(F^2-1)$ and $\pi(f)[\gamma, F]$ are compact operators on $\mathcal{H}$ for any $f \in C_0(X)$ and any $\gamma \in \Gamma$.

For $i = 0$, we require moreover $\mathcal{H}$ to be a $\mathbb{Z}/2$-graded Hilbert space, the representations of Γ and $C_0(X)$ to be by degree 0 operators (i.e. they preserve the grading), and F to be a degree 1 operator (i.e. it reverses the grading).

Clearly, by compressing to the orthogonal of the null space of $\pi(C_0(X))$, we may assume that π is a non-degenerate representation. Because the action of Γ on X is proper and X is Γ-compact, we shall see that we may actually assume, in the definition of a Kasparov element $(\mathcal{H}, \pi, F) \in K_i^\Gamma(X)$, that F is Γ-equivariant (i.e. $[\gamma, F] = 0$ for any $\gamma \in \Gamma$) and *properly supported*, i.e. for any $f \in C_c(X)$ there exists $g \in C_c(X)$ with $(\pi(g) - 1)F\pi(f) = 0$. The latter condition can be understood as a locality condition.

It turns out that properness of the action and Γ-compactness of X allow for the possibility of averaging over the group Γ, in order to make F equivariant. This is explained in the next lemma.

Lemma 2.5. *Fix a real-valued $f \in C_c(X)$. For $T \in \mathcal{L}(H)$, set*

$$A_f(T) = \sum_{\gamma\in\Gamma} \gamma\pi(f)T\pi(f)\gamma^{-1}.$$

Then

1. *The sum $\sum_{\gamma\in\Gamma} \gamma\pi(f)T\pi(f)\gamma^{-1}$ converges in the strong topology, and $A_f(T)$ is a bounded operator on $\mathcal{H}$; more precisely, there exists a constant $C > 0$, only depending on f, such that $\|A_f(T)\| \leq C\|T\|$.*
2. *$A_f(T)$ is properly supported (and, in particular, it maps $\pi(C_c(X))\mathcal{H}$ into itself).*
3. *$A_f(T)$ is Γ-equivariant, i.e. it commutes with Γ.*

Proof.
1. (Inspired by the proof of Lemma 3.2 in [Kas88].) Separating the real and imaginary parts of T, we may assume that T is self-adjoint. Then one has the operator inequalities
$$-\gamma\pi(f^2)\gamma^{-1}\|T\| \leq \gamma\pi(f)T\pi(f)\gamma^{-1} \leq \gamma\pi(f^2)\gamma^{-1}\|T\|$$
(for $\gamma \in \Gamma$). Summing over γ, one gets
$$-(\sum_{\gamma\in\Gamma} \pi(\gamma(f^2)))\|T\| \leq A_f(T) \leq \sum_{\gamma\in\Gamma} \pi(\gamma(f^2))\|T\|.$$
Thus $\|A_f(T)\| \leq \|\sum_{\gamma\in\Gamma} \gamma(f^2)\|_\infty \cdot \|T\|$.
2. For $g \in C_c(X)$, we have
$$A_f(T)\pi(g) = \sum_{\gamma\in\Gamma} \pi(\gamma(f))\gamma F\pi(f.\gamma^{-1}(g))\gamma^{-1}.$$
Consider the finite set $F = \{\gamma \in \Gamma : f.\gamma^{-1}(g) \neq 0\}$, and choose $h \in C_c(X)$ equal to 1 on $\bigcup_{\gamma\in F} \gamma(supp(f))$. Then $\pi(h)A_f(T)\pi(g) = A_f(T)\pi(g)$, which shows that $A_f(T)$ is properly supported.
3. Obvious, remembering that Γ acts on $\mathcal{L}(H)$ by the adjoint action: $\gamma(T) = \gamma T\gamma^{-1}$.

□

We say that two Kasparov triples are *operatorially homotopic* if they are related by a norm-continuous deformation of the operator F, keeping fixed the underlying Hilbert space and the representations involved.

Proposition 2.6. *Let X be a proper and Γ-compact space. Any Kasparov triple $(\mathcal{H}, \pi, F)$ in $K_i^\Gamma(X)$ is operatorially homotopic to a Kasparov triple $(\mathcal{H}, \pi, G)$, where G is properly supported and Γ-equivariant.*

Proof. Since the space X is proper and Γ-compact, there exists a non-negative function $c \in C_c(X)$ such that $\sum_{\gamma\in\Gamma} c(\gamma x) = 1$ for every $x \in X$. Taking

$$h = \sqrt{c}, \tag{2.1}$$

we form $G = A_h(F)$.

It already follows from Lemma 2.5 that G is properly supported and Γ-equivariant. Let us check that, for any $f \in C_c(X)$, the operator $\pi(f)(F-G)$ is compact. But since

$$\sum_{\gamma\in\Gamma} \gamma\pi(h^2)\gamma^{-1} = 1 \tag{2.2}$$

(as a consequence of the fact that π is non-degenerate), we have

$$\begin{aligned} F-G &= \textstyle\sum_{\gamma\in\Gamma}(\gamma\pi(h^2)\gamma^{-1}F - \gamma\pi(h)F\pi(h)\gamma^{-1}) \\ &= \textstyle\sum_{\gamma\in\Gamma}\gamma\pi(h)[\pi(h)\gamma^{-1},F] \\ &= \textstyle\sum_{\gamma\in\Gamma}\gamma\pi(h)(\pi(h)[\gamma^{-1},F]+[\pi(h),F]\gamma^{-1}). \end{aligned}$$

We remark that, in the last summation, all terms are compact, thanks to the assumptions on F. Then

$$\pi(f)(F-G) = \sum_{\gamma\in\Gamma}\gamma\pi(\gamma^{-1}(f).h)(\pi(h)[\gamma^{-1},F]+[\pi(h),F]\gamma^{-1}).$$

Since there are finitely many γ's such that $\gamma^{-1}(f).h$ is non-zero, we see that $\pi(f)(F-G)$ is compact, as a finite sum of compact operators. Therefore G defines a Kasparov triple which is operatorially homotopic to F, via the homotopy $F_t = (1-t)G+tF$ $(t\in[0,1])$. □

From now on, we shall assume that all operators defining Kasparov triples are Γ-equivariant and properly supported.

It is clear that the direct sum of two Kasparov triples over X is again a Kasparov triple over X. The equivalence relation which turns the set of Kasparov triples over X into the group $K_i^\Gamma(X)$ is the one described in [Kas95]; actually $K_i^\Gamma(X)$ is the Kasparov group $KK_i^\Gamma(C_0(X),\mathbb{C})$.

2.3. Definition of the assembly maps

We proceed to define the *analytic assembly map*

$$\tilde{\mu}_i^\Gamma : RK_i^\Gamma(\underline{E}\Gamma) \to K_i(C^*\Gamma),$$

by suitably modifying the construction given in [BCH94], (3.8), for the map μ_i^Γ : $RK_i^\Gamma(\underline{E}\Gamma) \to K_i(C_r^*\Gamma)$.

We begin with a locally compact, proper Γ-space X and a Hilbert space $\mathcal{H}$ endowed with a covariant representation π of $C_0(X)$. Consider the Γ-module $\pi(C_c(X))\mathcal{H}$ (since π is non-degenerate, this is a dense subspace of $\mathcal{H}$); view it as a right $\mathbb{C}\Gamma$-module by

$$\xi\cdot\gamma = \gamma^{-1}\xi; \tag{2.3}$$

define a $\mathbb{C}\Gamma$-valued scalar product on that module by

$$\langle\xi_1|\xi_2\rangle(\gamma) = \langle\xi_1\cdot\gamma|\xi_2\rangle = \langle\xi_1|\gamma\xi_2\rangle \tag{2.4}$$

(that $\gamma \mapsto \langle \xi_1 | \xi_2 \rangle(\gamma)$ has finite support in Γ already uses properness of the Γ-action) [4].

Since our aim is to complete $\pi(C_c(X))\mathcal{H}$ as a Hilbert C*-module, we have to discuss positivity of the functions $\gamma \mapsto \langle \xi | \gamma\xi \rangle$ (with $\xi \in \pi(C_c(X))\mathcal{H}$) as elements either in the reduced or the full C*-algebra of Γ. Recall that a function ϕ on Γ is *positive-definite* if it is of the form $\phi(g) = \langle \eta | \rho(g)\eta \rangle$ for some unitary representation ρ of Γ and some vector η in the Hilbert space of ρ.

For the reduced C*-algebra $C_r^*\Gamma$, an element $\gamma \mapsto \langle \xi | \gamma\xi \rangle$ as above, is positive for an abstract reason: any finitely supported, positive definite function on Γ defines a positive element in $C_r^*\Gamma$, by [Dix77], 13.7.8. If Γ is amenable, then $C_r^*\Gamma = C^*\Gamma$, so that trivially the same elements are positive in $C^*\Gamma$. But this abstract argument fails in general for the full C*-algebra: it turns out that, if every finitely supported, positive definite function on Γ is a positive element in $C^*\Gamma$, then Γ is amenable (see [Val98]).

Lemma 2.7. *Let X be a locally compact, proper Γ-space; let π be a Γ-covariant representation of $C_0(X)$ on a Hilbert space $\mathcal{H}$. For any $\xi \in \pi(C_c(X))\mathcal{H}$, the function $\gamma \mapsto \langle \xi | \gamma\xi \rangle$ defines a positive element in $C^*\Gamma$.*

Before proving Lemma 2.7, we introduce some preliminary notations that will be important for the sequel.

Consider the space $C_c(\Gamma, \mathcal{H})$ of $\mathcal{H}$-valued, finitely supported functions on Γ; it is viewed as a right $\mathbb{C}\Gamma$-module via ordinary convolution:

$$\xi a(\sigma) = \sum_{\gamma \in \Gamma} \xi(\gamma) a(\gamma^{-1}\sigma)$$

($\xi \in C_c(\Gamma, \mathcal{H})$, $a \in \mathbb{C}\Gamma$, $\sigma \in \Gamma$; notice that this does not involve the left action of Γ on $\mathcal{H}$). The module $C_c(\Gamma, \mathcal{H})$ carries a $\mathbb{C}\Gamma$-valued scalar product:

$$\langle \xi | \eta \rangle(\sigma) = \sum_{\gamma \in \Gamma} \langle \xi(\gamma) | \eta(\gamma\sigma) \rangle \tag{2.5}$$

(ξ, $\eta \in C_c(\Gamma, \mathcal{H})$, $\sigma \in \Gamma$) which is positive, i.e. $\langle \xi | \xi \rangle$ is a positive element in $C^*\Gamma$. To see the latter, choose a basis $(e_i)_{i \in I}$ of $\mathcal{H}$, and write $\xi(\sigma) = \sum_{i \in I} \xi_i(\sigma) e_i$ (so that ξ_i is in $\mathbb{C}\Gamma$). Then

$$\langle \xi | \xi \rangle(\sigma) = \sum_{\gamma \in \Gamma} \sum_{i \in I} \overline{\xi_i(\gamma)} \xi_i(\gamma\sigma) = \sum_{i \in I} \sum_{\gamma \in \Gamma} \xi_i^*(\gamma) \xi_i(\gamma^{-1}\sigma),$$

i.e. $\langle \xi | \xi \rangle = \sum_{i \in I} \xi_i^* \xi_i$, which is a positive element in $C^*\Gamma$. This also shows that the completion of $C_c(\Gamma, \mathcal{H})$ as a $C^*\Gamma$-module is the standard module

$$\mathcal{H} \otimes C^*\Gamma = \{(\xi_i)_{i \in I} : \xi_i \in C^*\Gamma, \sum_{i \in I} \xi_i^* \xi_i \textit{ converges in } C^*\Gamma\}.$$

[4] Our convention for scalar products: linear on the right, anti-linear on the left.

Proof of Lemma 2.7. The proof of this lemma (and also of the following one) uses ideas from [Pie00], Proposition 2.3.2.

Let $h \in C_c(X)$ be as in formula (2.1), i.e. $h \geq 0$ and $\sum_{\gamma\in\Gamma} \gamma h^2 = 1$. Let $\ell^2(\Gamma, \mathcal{H})$ be the Hilbert space of $\mathcal{H}$-valued square-summable functions on Γ. Define

$$S : \begin{cases} \mathcal{H} & \rightarrow \quad \ell^2(\Gamma, \mathcal{H}) \\ \xi & \mapsto \quad (\gamma \mapsto \pi(h)\gamma\xi) \end{cases}$$

Note that $S\xi$ really belongs to $\ell^2(\Gamma, \mathcal{H})$, since

$$\|S\xi\|^2 = \sum_{\gamma\in\Gamma} \|\gamma\pi(\gamma^{-1}h)\xi\|^2 = \sum_{\gamma\in\Gamma} \|\pi(\gamma^{-1}h)\xi\|^2$$

$$= \sum_{\gamma\in\Gamma} \langle \pi(\gamma^{-1}h)^2\xi|\xi\rangle = \|\xi\|^2.$$

It is clear that S maps $\pi(C_c(X))\mathcal{H}$ into $C_c(\Gamma, \mathcal{H})$, and it is easy to check that S commutes with the right Γ-actions on these two spaces.

Now define an operator

$$S^* : \begin{cases} C_c(\Gamma, \mathcal{H}) & \rightarrow \quad \pi(C_c(X))\mathcal{H} \\ \eta & \mapsto \quad \sum_{t\in\Gamma} t^{-1}\pi(h)(\eta(t)) \end{cases}$$

(the notation S^* will be justified in a minute). For the moment, notice that S^*S is the identity on $\pi(C_c(X))\mathcal{H}$. Now for $\xi \in \pi(C_c(X))\mathcal{H}$, $\eta \in C_c(\Gamma, \mathcal{H})$, $\gamma \in \Gamma$:

$$\langle \xi|S^*\eta\rangle(\gamma) = \langle \xi|\gamma S^*\eta\rangle = \sum_{t\in\Gamma} \langle \xi|\gamma t^{-1}\pi(h)(\eta(t))\rangle$$

$$= \sum_{s\in\Gamma} \langle \xi|s^{-1}\pi(h)(\eta(s\gamma))\rangle = \sum_{s\in\Gamma} \langle \pi(h)s\xi|\eta(s\gamma)\rangle$$

$$= \sum_{s\in\Gamma} \langle s\pi(s^{-1}h)\xi|\eta(s\gamma)\rangle = \sum_{s\in\Gamma} \langle S\xi(s)|\eta(s\gamma)\rangle = \langle S\xi|\eta\rangle(\gamma).$$

This proves that S and S^* are really adjoint to each other, with respect to the $\mathbb{C}\Gamma$-valued scalar products. In particular, for $\xi \in \pi(C_c(X))\mathcal{H}$, we have:

$$\langle \xi|\xi\rangle(\cdot) = \langle \xi|S^*S\xi\rangle(\cdot) = \langle S\xi|S\xi\rangle(\cdot).$$

But we have seen, just before this proof, that $\langle S\xi|S\xi\rangle(\cdot)$ is a positive element in $C^*\Gamma$. Therefore, so is $\langle \xi|\xi\rangle(\cdot)$. □

By Lemma 2.7, we may form the completion of $\pi(C_c(X))\mathcal{H}$ with respect to the scalar product given by formula (2.4), and get a Hilbert C*-module $\mathcal{E}$ over $C^*\Gamma$.

Lemma 2.8. *Let $T \in \mathcal{L}(H)$ be properly supported and such that $T\gamma = \gamma T$ for any $\gamma \in \Gamma$. Then T extends continuously to an operator $\mathcal{T} \in \mathcal{L}_{C^*\Gamma}(\mathcal{E})$.*

Proof. Let S, S^* be the operators appearing in the proof of Lemma 2.7, such that $S^*S = 1$. For $\eta \in C_c(\Gamma, \mathcal{H})$, $\gamma \in \Gamma$, an easy computation shows

$$STS^*(\eta)(\gamma) = \sum_{s\in\Gamma} \pi(h)T\pi(sh)s\eta(s^{-1}\gamma).$$

If we identify $C_c(\Gamma, \mathcal{H})$ with the algebraic tensor product $\mathcal{H} \otimes \mathbb{C}\Gamma$, this can be re-written

$$STS^* = \sum_{s\in\Gamma} \pi(h)T\pi(sh)s \otimes \lambda_\Gamma(s)$$

where λ_Γ denotes the left regular representation of Γ. Since T is properly supported, this sum is finite. It is then clear that STS^* extends to a continuous $C^*\Gamma$-module map on $\mathcal{H} \otimes C^*\Gamma$. So $T = S^*(STS^*)S$ extends to a continuous operator $\mathcal{T} \in \mathcal{L}_{C^*\Gamma}(\mathcal{E})$. □

Lemmas 2.7 and 2.8 are much easier to prove when $C^*\Gamma$ is replaced by $C_r^*\Gamma$ (see Lemma 6.1.3 in [Val02]).

Let then X be a Γ-compact subset of $\underline{E}\Gamma$, and let $(\mathcal{H}, \pi, F)$ be an element of $KK_i^\Gamma(C_0(X), \mathbb{C})$, where F is Γ-equivariant and properly supported. As above, let $\mathcal{E}$ be the completion of $\pi(C_c(X))\mathcal{H}$ as a Hilbert C*-module over $C^*\Gamma$. The operator F satisfies the assumptions of Lemma 2.8, so it extends to an operator $\mathcal{F} \in \mathcal{L}_{C^*\Gamma}(\mathcal{E})$. Assume for a moment that the following result (proved in Section 2.4) is true.

Proposition 2.9. *$\mathcal{F}^2 - 1$ is a compact operator on the C^*-module $\mathcal{E}$.*

This proposition says that the pair $(\mathcal{E}, F)$ defines an element in

$$KK_i(\mathbb{C}, C^*\Gamma) = K_i(C^*\Gamma).$$

In (3.10) of [BCH94], this element is called the *Γ-index* of F and denoted by $Index_\Gamma(F)$. It is immediate that the homomorphism

$$KK_i^\Gamma(C_0(X), \mathbb{C}) \to K_i(C^*\Gamma)$$

extends to the direct limit $RK_i^\Gamma(\underline{E}\Gamma)$ of Definition 2.4.

Definition 2.10. The homomorphism

$$\tilde{\mu}_i^\Gamma : \begin{cases} RK_i^\Gamma(\underline{E}\Gamma) \to K_i(C^*\Gamma) \\ (\mathcal{H}, \pi, F) \mapsto Index_\Gamma(F) \end{cases}$$

is the *analytical assembly map.*

We also *define* the analytical assembly map $\mu_i^\Gamma : RK_i^\Gamma(\underline{E}\Gamma) \to K_i(C_r^*\Gamma)$ by $\mu_i^\Gamma = (\lambda_\Gamma)_* \circ \tilde{\mu}_i^\Gamma$. We remark that the map $\tilde{\mu}_i^\Gamma$ is denoted by β in [Kas75], [Kas95], [Kas88], [Ros83]; and by A in [FRR95]. The map μ_i^Γ is denoted by A' in [FRR95]. The reader may also enjoy Gromov's point of view in [Gro93], where the analytical assembly map is denoted by K.

Example 2.11. Let H be a finite subgroup of Γ. Fix a finite-dimensional unitary representation ρ of H on a complex vector space V_ρ. Let us describe an element $\beta_{H,\rho}$ of $RK_0^\Gamma(\underline{E}\Gamma)$ as follows: take $X_H = \Gamma/H$, with action of Γ by left translations; X_H is a proper, Γ-compact space. In the picture of $\underline{E}\Gamma$ by probability measures on Γ, the space X_H identifies with the set of uniform probability measures on left cosets of H (in particular, for $H = 1$, the space X_H identifies with the set of Dirac measures). Take now the induced vector bundle $E_\rho = \Gamma \times_H V_\rho$ over X_H; denote by $\mathcal{H}$ the space of ℓ^2-sections of E_ρ; this is nothing but the space of the representation obtained by inducing up ρ from H to Γ. Consider the Γ-covariant representation π of $C_0(X_H)$ by pointwise multiplication of sections; since X_H is discrete and the fibers of E_ρ are of finite dimension, the representation π acts by compact operators on $\mathcal{H}$, so the triple $(\mathcal{H}, \pi, 0)$ defines an element $\beta_{H,\rho}$ of $RK_0^\Gamma(\underline{E}\Gamma)$. To describe the image $\mu_0^\Gamma(\beta_{H,\rho})$, we may clearly assume that ρ is irreducible. Then

$$\pi(C_c(X_H))\mathcal{H} = \mathbb{C}\Gamma \bigotimes_{\mathbb{C}H} V_\rho,$$

the space of finitely supported sections of E_ρ. Since ρ is irreducible, there exists a minimal projection $p_{H,\rho} \in \mathbb{C}H$ such that the $\mathbb{C}H$-modules V_ρ and $\mathbb{C}H \star \overline{p_{H,\rho}}$ are isomorphic (where $\mathbb{C}H \star \overline{p_{H,\rho}}$ is the left ideal in $\mathbb{C}H$ generated by the complex conjugate of $p_{H,\rho}$ - recall that the product in $\mathbb{C}H$ is given by convolution, and is denoted by $\star$). It is worthwhile to spell out this isomorphism explicitly. Let ξ be a vector of norm 1 in V_ρ. Then

$$p_{H,\rho}(s) = \frac{\deg \rho}{|H|} \langle \rho(s)\xi | \xi \rangle$$

for $s \in H$ (it follows from the Schur orthogonality relations, see [Dix77] 14.3.3, that $p_{H,\rho}$ is indeed a projection in $\mathbb{C}H$). This projection $p_{H,\rho}$ is characterized by the fact that $\sigma(p_{H,\rho}) = 0$ for every irreducible representation σ of H which is not equivalent to $\overline{\rho}$, the contragredient representation of ρ; while $\overline{\rho}(p_{H,\rho})$ is the orthogonal projection on the one-dimensional subspace $\mathbb{C}\xi$ of $V_{\overline{\rho}}$. The map

$$\mathbb{C}H \star \overline{p_{H,\rho}} \to V_\rho : f \mapsto \rho(f)\xi$$

is an isomorphism intertwining the left regular representation λ_H (restricted to $\mathbb{C}H \star \overline{p_{H,\rho}}$) and ρ. Then, for $\eta \in \mathbb{C}\Gamma$, set

$$\check{\eta}(\gamma) = \eta(\gamma^{-1})$$

($\gamma \in \Gamma$). Recall that Γ acts on the right on $\mathcal{H}$ by

$$\eta \cdot \gamma = \gamma^{-1}\eta$$

while it acts on the right on $\mathbb{C}\Gamma$ by right convolution by Dirac measures. Using the fact that $\check{\overline{p}}_{H,\rho} = p_{H,\rho}^* = p_{H,\rho}$, it is easy to check that the map

$$\Psi : \mathbb{C}\Gamma \bigotimes_{\mathbb{C}H} (\mathbb{C}H \star \overline{p_{H,\rho}}) \to p_{H,\rho} \star \mathbb{C}\Gamma : a \otimes b \mapsto \check{b} \star \check{a}$$

is a map of right $\mathbb{C}\Gamma$-modules, which moreover preserves the $C^*\Gamma$-valued scalar products, i.e.

$$\langle a_1 \otimes b_1 | a_2 \otimes b_2 \rangle_{C^*\Gamma} = (\check{b_1} \star \check{a_1})^* \star (\check{b_2} \star \check{a_2})$$

($a_1, a_2 \in \mathbb{C}\Gamma$, $b_1, b_2 \in \mathbb{C}H \star p_{H,\rho}$). So Ψ extends to an isometry of Hilbert $C^*\Gamma$-modules, from the completion of $\mathbb{C}\Gamma \bigotimes_{\mathbb{C}H} (\mathbb{C}H \star \overline{p_{H,\rho}})$ to the right ideal $p_{H,\rho} C^*\Gamma$ in $C^*\Gamma$. In a more down-to-earth language, $\tilde{\mu}_0^\Gamma(\beta_{H,\rho})$ is just the class $[p_{H,\rho}]$ of the projection $p_{H,\rho}$ in $K_0(C^*\Gamma)$.

In particular for $H = 1$ and $\rho = 1_H$ the trivial one-dimensional representation, we have $\tilde{\mu}_0^\Gamma(\beta_{H,1_H}) = [1]$, the K-theory class of the unit of $C^*\Gamma$. This element has infinite order in $K_0(C^*\Gamma)$, since it maps to 1 under the trivial one-dimensional representation of Γ. Its image $\mu_0^\Gamma(\beta_{H,1_H}) = [1]$ in $K_0(C_r^*\Gamma)$ also has infinite order, since the canonical trace on $C_r^*\Gamma$ maps [1] to 1. This can be rephrased by saying that there is a commutative diagram

$$\begin{array}{ccc} RK_0^\Gamma(\underline{E}\Gamma) & \overset{\mu_0^\Gamma}{\rightarrow} & K_0(C_r^*\Gamma) \\ \uparrow & & \uparrow \\ H_0(\Gamma, \mathbb{Z}) & = & \mathbb{Z} \end{array}$$

where vertical maps are monomorphisms.

Suppose that Γ is torsion-free; then

$$RK_0^\Gamma(\underline{E}\Gamma) = RK_0^\Gamma(E\Gamma) = RK_0(B\Gamma);$$

also, as mentioned in the Introduction, one has $RK_0(B\Gamma) \bigotimes_{\mathbb{Z}} \mathbb{C} = \bigoplus_{n=0}^{\infty} H_{2n}(\Gamma, \mathbb{C})$. So the preceding diagram shows that the Baum–Connes conjecture holds on the 0-dimensional part of $RK_0^\Gamma(E\Gamma)$.

Suppose at the other extreme that Γ is a finite group. In this case both $K_0^\Gamma(pt)$ and $K_0(C_r^*\Gamma)$ are abstractly isomorphic to the additive group of the representation ring $R(\Gamma)$, i.e. to the free abelian group on the set $\hat{\Gamma}$ of isomorphism classes of irreducible representations of Γ. In the above construction, take $H = \Gamma$ and let ρ run along $\hat{\Gamma}$. Then the $\beta_{\Gamma,\rho}$'s run along a set of generators of $K_0^\Gamma(pt)$ and the $[p_{\Gamma,\rho}]$'s run along a set of generators of $K_0(C_r^*\Gamma)$. On the other hand:

$$RK_1^\Gamma(\underline{E}\Gamma) = 0 = K_1(C_r^*\Gamma)$$

for Γ finite. In other words, we have checked that the Baum–Connes conjecture holds for finite groups.

2.4. Equivalence of two definitions

Here we shall simultaneously prove Proposition 2.9 and give an alternative construction of the analytical assembly map. First, we take a closer look at the situation described in Lemma 2.5, i.e. a proper Γ-space X and a Hilbert space $\mathcal{H}$ endowed with a covariant representation π of $C_0(X)$. The next lemma complements Lemma 2.5.

Lemma 2.12. *Fix a real-valued* $f \in C_c(X)$. *For* $T \in \mathcal{L}(H)$, *set*

$$A_f(T) = \sum_{\gamma \in \Gamma} \gamma \pi(f) T \pi(f) \gamma^{-1}$$

as in Lemma 2.5. Then

1. $A_f(T)$ *extends continuously to an operator* $\mathcal{A}_f(T) \in \mathcal{L}_{C^*\Gamma}(\mathcal{E})$.
2. *If* T *is a compact operator on* $\mathcal{H}$, *then* $\mathcal{A}_f(T)$ *is a compact operator on* $\mathcal{E}$.

Proof. 1. Lemma 2.5 shows that $A_f(T)$ is Γ-equivariant and properly supported. So Lemma 2.8 applies.

2. Consider again the function h of formula (2.1), and the operators S, S^* appearing in the proof of Lemma 2.7. In the proof of Lemma 2.8, we saw that $STS^* = \sum_{s \in \Gamma} \pi(h) T \pi(sh) s \otimes \lambda_\Gamma(s)$, provided T is Γ-equivariant, properly supported on $\mathcal{H}$. So replacing T by $A_f(T)$, we get

$$SA_f(T)S^* = \sum_{s \in \Gamma} \sum_{\gamma \in \Gamma} \pi(h.\gamma f) \gamma T \pi(f.\gamma^{-1} sh) \gamma^{-1} s \otimes \lambda_\Gamma(s).$$

This double sum is finite, as a consequence of properness of the Γ-action on X. If T is compact on $\mathcal{H}$, then $SA_f(T)S^*$ extends to an operator on $\mathcal{H} \otimes C^*\Gamma$, which belongs to $\mathcal{K}(H) \otimes C^*\Gamma = \mathcal{K}(H \otimes C^*\Gamma)$. So $\mathcal{A}_f(T) = S^*(SA_f(T)S^*)S$ extends to a compact operator $\mathcal{A}_f(T)$ on $\mathcal{E}$.

□

We observe now that, if X is a proper, Γ-compact space, then there is a canonical element $[\mathcal{L}_X]$ in the K_0-group of the crossed product $C_0(X) \rtimes \Gamma$.

Recall from [Ped79], 7.6.5, that $C_0(X) \rtimes \Gamma$ is the universal C*-completion of $C_c(X \times \Gamma)$, with convolution product given by

$$f_1 \star f_2(x, \sigma) = \sum_{\gamma \in \Gamma} f_1(x, \gamma) f_2(\gamma^{-1}x, \gamma^{-1}\sigma)$$

(f_1, $f_2 \in C_c(X \times \Gamma)$, $x \in X$, $\sigma \in \Gamma$), and involution given by

$$f^*(x, \sigma) = \overline{f(\sigma^{-1}x, \sigma^{-1})}$$

($f \in C_c(X \times \Gamma)$, $x \in X$, $\sigma \in \Gamma$). Alternately, $C_0(X) \rtimes \Gamma$ can be viewed as the universal C*-completion of the space of finite, formal sums $\sum_{\gamma \in \Gamma} f_\gamma.\gamma$, with $f_\gamma \in C_0(X)$. The product is:

$$\left(\sum_{s \in \Gamma} f_s.s\right)\left(\sum_{t \in \Gamma} g_t.t\right) = \sum_{\gamma \in \Gamma}\left(\sum_{s \in \Gamma} f_s s(g_{s^{-1}\gamma})\right).\gamma,$$

and the involution is:

$$(\sum_{\gamma\in\Gamma} f_\gamma.\gamma)^* = \sum_{\gamma\in\Gamma} \gamma(\overline{f_{\gamma^{-1}}}).\gamma.$$

Now we give the construction of the element $[\mathcal{L}_X]$. We first view $C_c(X)$ as a right $C_c(X \times \Gamma)$-module via the formula

$$\eta f(x) = \sum_{\gamma\in\Gamma} \eta(\gamma x) f(\gamma x, \gamma) \tag{2.6}$$

($\eta \in C_c(X)$, $f \in C_c(X \times \Gamma), x \in X$). Moreover the formula

$$\langle \xi|\eta\rangle(x,\gamma) = \overline{\xi(x)}\eta(\gamma^{-1}x) \tag{2.7}$$

(ξ, $\eta \in C_c(X)$, $x \in X$, $\gamma \in \Gamma$) defines a $C_c(X\times\Gamma)$-valued scalar product on $C_c(X)$, and the following lemma shows that this scalar product is indeed positive-definite. Note that we may also write

$$\langle \xi|\eta\rangle = \sum_{\gamma\in\Gamma} \bar{\xi}\gamma(\eta).\gamma.$$

Lemma 2.13. *For $\eta \in C_c(X)$, the element $\langle\eta|\eta\rangle$ is positive in the crossed product C^*-algebra $C_0(X) \rtimes \Gamma$.*

Proof. Since the Γ-action on X is proper, hence also amenable, the full crossed product $C_0(X) \rtimes \Gamma$ is isomorphic to the reduced crossed product $C_0(X) \rtimes_r \Gamma$, by [AD87]. Let us describe the latter. Consider the trivial field of Hilbert spaces $X \times \ell^2(\Gamma)$ over X: its set of continuous sections is $C_0(X, \ell^2(\Gamma))$, a Hilbert C*-module over $C_0(X)$ (with $C_0(X)$-valued scalar product given by pointwise scalar product of sections); we view sections in $C_0(X, \ell^2(\Gamma))$ as functions on $X \times \Gamma$. We define a *-representation of $C_c(X \times \Gamma)$ on $C_0(X, \ell^2(\Gamma))$ by

$$f\xi(x,\gamma) = \sum_{\sigma\in\Gamma} f(\gamma x, \sigma).\xi(x, \sigma^{-1}\gamma)$$

($f \in C_c(X \times \Gamma)$, $\xi \in C_0(X, \ell^2(\Gamma))$, $x \in X$, $\gamma \in \Gamma$). It is a known fact (see e.g. [Kas88], 3.7) that this *-homomorphism extends to a faithful *-representation of $C_0(X) \rtimes_r \Gamma$ on $C_0(X, \ell^2(\Gamma))$. For $\eta \in C_c(X)$, we then have to prove that $\langle\eta|\eta\rangle$ is a positive element in the C*-algebra $\mathcal{L}_{C_0(X)}(C_0(X, \ell^2(\Gamma)))$. By Lemma 4.1 in [Lan95], this amounts to proving that, for every $\xi \in C_0(X, \ell^2(\Gamma))$, the scalar product $\langle\langle\eta|\eta\rangle\xi|\xi\rangle$ is a positive element in the C*-algebra $C_0(X)$, i.e. is a non-negative function on X. So, for $x \in X$, we compute:

$$\begin{aligned}
\langle\langle\eta|\eta\rangle\xi|\xi\rangle(x) &= \sum_{\gamma\in\Gamma} \overline{(\langle\eta|\eta\rangle\xi)(x,\gamma)}\xi(x,\gamma) \\
&= \sum_{\gamma\in\Gamma}\sum_{\sigma\in\Gamma} \overline{\langle\eta|\eta\rangle(\gamma x,\sigma)\xi(x,\sigma^{-1}\gamma)}\xi(x,\gamma) \\
&= \sum_{\gamma\in\Gamma}\sum_{\sigma\in\Gamma} \eta(\gamma x)\xi(x,\gamma)\overline{\eta(\sigma^{-1}\gamma x)\xi(x,\sigma^{-1}\gamma)} \\
&= \sum_{\gamma\in\Gamma} \eta(\gamma x)\xi(x,\gamma)\cdot\sum_{\sigma\in\Gamma}\overline{\eta(\sigma x)\xi(x,\sigma)} \\
&= \left|\sum_{\gamma\in\Gamma}\eta(\gamma x)\xi(x,\gamma)\right|^2 \geq 0.
\end{aligned}$$

This conclues the proof of the lemma. □

Form the completion of $C_c(X)$ with respect to the scalar product defined by (2.7), and get a C*-module $\mathcal{L}_X$ over $C_0(X) \rtimes \Gamma$. We recall that a *rank 1* operator on a C*-module $\mathcal{E}$ is an operator of the form $\xi \mapsto \theta_{\phi,\psi}(\xi)) = \phi\langle\psi|\xi\rangle$, for fixed $\phi, \psi \in \mathcal{E}$.

Lemma 2.14. *The identity of $\mathcal{L}_X$ is a rank* 1 *operator, in the sense of C*-modules.*

Proof. For $\xi, \phi, \psi \in C_c(X)$ and $x \in X$, we get, combining (2.6) and (2.7):

$$(\theta_{\phi,\psi}(\xi))(x) = \sum_{\gamma\in\Gamma}\phi(\gamma x)\langle\psi|\xi\rangle(\gamma x,\gamma) = (\sum_{\gamma\in\Gamma}\phi(\gamma x)\overline{\psi(\gamma x)})\xi(x).$$

Let $h \in C_c(X)$ be defined as in equation (2.1); since $\sum_{\gamma\in\Gamma} h^2(\gamma x) = 1$ for every $x \in X$, we see that $\theta_{h,h}$ is the identity on $\mathcal{L}_X$. □

This lemma shows that $\mathcal{L}_X$ defines an element $[\mathcal{L}_X]$ of $K_0(C_0(X) \rtimes \Gamma)$. Notice that since $h\langle h|\xi\rangle = \xi$ for any $\xi \in C_c(X)$, we get $\langle h|h\rangle \star \langle h|\xi\rangle = \langle h|\xi\rangle$ and in particular $\langle h|h\rangle^2 = \langle h|h\rangle$, so that $p = \langle h|h\rangle$ is a projector in $C_c(X \times \Gamma)$. The map $p \star C_c(X \times \Gamma) \to C_c(X) : p \star f \mapsto h \cdot f$ is then well-defined, and identifies the right ideal $p.(C_0(X) \rtimes \Gamma)$ of $C_0(X) \rtimes \Gamma$ with the C*-module $\mathcal{L}_X$.

We shall need the *descent homomorphism*

$$j_\Gamma : KK_i^\Gamma(C_0(X), \mathbb{C}) \to KK_i(C_0(X) \rtimes \Gamma, C^*\Gamma),$$

also known as *induction to the crossed product*, see [Kas95], [Kas88]. For $x = (\mathcal{H}, \pi, F) \in KK_i^\Gamma(C_0(X), \mathbb{C})$, the element $j_\Gamma(x)$ is described as follows. Recall that we constructed a Hilbert $C^*\Gamma$-module $\tilde{\mathcal{H}} = \mathcal{H} \otimes C^*\Gamma$ as the completion of $C_c(\Gamma, \mathcal{H})$ for the $\mathbb{C}\Gamma$-valued scalar product in formula (2.5). There is an isometric left Γ-action on $C_c(\Gamma, \mathcal{H})$ given by

$$(\gamma \cdot \xi)(\sigma) = \gamma(\xi(\gamma^{-1}\sigma))$$

($\xi \in C_c(\Gamma, \mathcal{H})$; $\gamma, \sigma \in \Gamma$); there is also a Γ-covariant left $C_0(X)$-action $\tilde{\pi}$ on $C_c(\Gamma, \mathcal{H})$ given by

$$(\tilde{\pi}(f) \cdot \xi)(\sigma) = \pi(f)(\xi(\sigma))$$

($\xi \in C_c(\Gamma, \mathcal{H})$, $f \in C_c(X)$, $\sigma \in \Gamma$). In the identification of $C_c(\Gamma, \mathcal{H})$ with $\mathcal{H} \otimes \mathbb{C}\Gamma$, this can be re-written $\tilde{\pi}(f) = \pi(f) \otimes 1$, so that it extends to a left action of $C_0(X)$ on $\tilde{\mathcal{H}}$; being Γ-covariant, this action extends to the crossed product $C_0(X) \rtimes \Gamma$; the "integrated" form of that action is:

$$(\tilde{\pi}(\sum_{\gamma \in \Gamma} f_\gamma \cdot \gamma)\xi)(\sigma) = \sum_{\gamma \in \Gamma} \pi(f_\gamma)\gamma(\xi(\gamma^{-1}\sigma))$$

($f_\gamma \in C_0(X)$, $\xi \in C_c(\Gamma, \mathcal{H})$, $\sigma \in \Gamma$). Finally, define the operator $\tilde{F}$ on $C_c(\Gamma, \mathcal{H})$ by

$$(\tilde{F}\xi)(\gamma) = F(\xi(\gamma))$$

($\xi \in C_c(\Gamma, \mathcal{H})$, $\gamma \in \Gamma$). The operator $\tilde{F}$ extends to an operator on $\tilde{\mathcal{H}}$, and the triple $(\tilde{\mathcal{H}}, \tilde{\pi}, \tilde{F})$ defines the Kasparov element $j_\Gamma(x) \in KK_i(C_0(X) \rtimes \Gamma, C^*\Gamma)$. Next we wish to perform the Kasparov product

$$[\mathcal{L}_X] \otimes_{C_0(X) \rtimes \Gamma} j_\Gamma(x) \in KK_i(\mathbb{C}, C^*\Gamma) = K_i(C^*\Gamma),$$

As above, we take $x = (\mathcal{H}, \pi, F)$ with F properly supported and Γ-equivariant; we denote by $\mathcal{E}$ the C*-module completion of $\pi(C_c(X))\mathcal{H}$, as in Lemma 2.8. We shall need the peculiar function $h \in C_c(X)$ appearing in equation (2.1).

Lemma 2.15. *The element* $[\mathcal{L}_X] \otimes_{C_0(X) \rtimes \Gamma} j_\Gamma(x)$ *can be represented by the pair* $(\mathcal{E}, A_h(F))$.

Proof. Since $\mathcal{L}_X$ can be described as the right ideal $p.C_0(X) \rtimes \Gamma$, or alternatively by the *-homomorphism

$$\alpha : \mathbb{C} \to C_0(X) \rtimes \Gamma : 1 \mapsto p,$$

the Kasparov product $[\mathcal{L}_X] \bigotimes_{C_0(X) \rtimes \Gamma} j_\Gamma(x) = \alpha^*(j_\Gamma(x))$ is represented by the triple $(\tilde{\mathcal{H}}, \tilde{\pi}(p), \tilde{F})$, where the action of $\mathbb{C}$ on $\tilde{\mathcal{H}}$ is via the projector $\tilde{\pi}(p)$. Define then a map:

$$\beta : \begin{cases} \tilde{\pi}(p) C_c(\Gamma, \mathcal{H}) & \to & \pi(C_c(X))\mathcal{H} \\ \tilde{\pi}(p)\xi & \mapsto & \sum_{\gamma \in \Gamma} \gamma^{-1}\pi(h)(\xi(\gamma)). \end{cases}$$

For ξ, $\eta \in C_c(\Gamma, \mathcal{H})$, the following relation holds in $\mathbb{C}\Gamma$:

$$\langle \tilde{\pi}(p)\xi | \tilde{\pi}(p)\eta \rangle = \langle \sum_{\gamma \in \Gamma} \gamma^{-1}\pi(h)(\xi(\gamma)) | \sum_{\gamma \in \Gamma} \gamma^{-1}\pi(h)(\eta(\gamma)) \rangle.$$

As a sample of the involved computations, let us check this equality; using equations (2.5), (2.4) and (2.2), we have, for $\sigma \in \Gamma$:

$$\begin{aligned}
\langle \tilde{\pi}(p)\xi | \tilde{\pi}(p)\eta\rangle(\sigma) &= \langle \tilde{\pi}(p)\xi|\eta\rangle(\sigma) \\
&= \sum_{\gamma\in\Gamma} \langle (\tilde{\pi}(p)\xi)(\gamma)|\eta(\gamma\sigma)\rangle \\
&= \sum_{\gamma\in\Gamma} \langle (\tilde{\pi}(\sum_{\delta\in\Gamma} h\delta(h).\delta)\xi)(\gamma)|\eta(\gamma\sigma)\rangle \\
&= \sum_{\gamma\in\Gamma}\sum_{\delta\in\Gamma} \langle \pi(h)\pi(\delta(h))\delta\xi(\delta^{-1}\gamma)|\eta(\gamma\sigma)\rangle \\
&= \sum_{\gamma\in\Gamma}\sum_{\delta\in\Gamma} \langle \delta\pi(h)\xi(\delta^{-1}\gamma)|\pi(h)\eta(\gamma\sigma)\rangle \\
&= \sum_{\gamma\in\Gamma}\sum_{\delta\in\Gamma} \langle \gamma\delta^{-1}\pi(h)\xi(\delta)|\pi(h)\eta(\gamma\sigma)\rangle \\
&= \langle \sum_{\delta\in\Gamma} \delta^{-1}\pi(h)(\xi(\delta))| \sum_{\gamma\in\Gamma} \gamma^{-1}\pi(h)(\eta(\gamma))\rangle(\sigma).
\end{aligned}$$

This shows that β is well-defined and extends to an isometric map of $C^*\Gamma$-modules between $\tilde{\pi}(p)\tilde{\mathcal{H}}$ and $\mathcal{E}$. Let us check that β is onto. For $f \in C_c(X)$, $\xi \in C_c(\Gamma, \mathcal{H})$, $\sigma \in \Gamma$, we have:

$$(\tilde{\pi}(\langle h|f\rangle)\xi)(\sigma) = \sum_{\gamma\in\Gamma} \pi(h)\pi(\gamma(f))\gamma(\xi(\gamma^{-1}\sigma)).$$

So that

$$\beta(\tilde{\pi}(p)(\tilde{\pi}(\langle h|f\rangle)\xi)) = \sum_{\gamma,\sigma\in\Gamma} \sigma^{-1}\pi(h^2)\gamma\pi(f)(\xi(\gamma^{-1}\sigma))$$

$$= \sum_{\gamma,\sigma\in\Gamma} \sigma^{-1}\pi(h^2)\sigma\gamma^{-1}\pi(f)(\xi(\gamma)) = \sum_{\gamma\in\Gamma} \gamma^{-1}\pi(f)(\xi(\gamma))$$

where the last equality again follows from (2.2). For $\eta \in \mathcal{H}$, define then a function $\overline{\eta} \in C_c(\Gamma, \mathcal{H})$ by

$$\overline{\eta} = \begin{cases} \eta & \text{if } \gamma = 1 \\ 0 & \text{if } \gamma \neq 1 \end{cases}$$

Then

$$\beta(\tilde{\pi}(p)(\tilde{\pi}(\langle h|f\rangle)\overline{\eta})) = \pi(f)\eta$$

shows that β is onto. We then transfer $\tilde{F}$ on $\mathcal{E}$ via β, and find by a simple computation

$$\beta\tilde{F}\beta^{-1} = \mathcal{A}_h(F).$$

□

Proof of Proposition 2.9. We have to show that $\mathcal{F}^2 - 1$ is a compact operator on the C*-module $\mathcal{E}$; on the other hand, we know by Lemma 2.15 that $(\mathcal{A}_h(F))^2 - 1$ is a compact operator on $\mathcal{E}$. So, to prove Proposition 2.9, it is enough to prove that $\mathcal{F} - \mathcal{A}_h(F)$ is compact on $\mathcal{E}$. We use the fact that F is properly supported; so we find $g \in C_c(X)$ such that $\pi(g)F\pi(h) = F\pi(h)$. Let $f \in C_c(X)$ be real-valued and equal to 1 on $supp(g) \cup supp(h)$, so that $f.g = g$ and $f.h = h$. Then, on $\pi(C_c(X))\mathcal{H}$, one has, using (2.2):

$$
\begin{aligned}
F - \mathcal{A}_h(F) &= \textstyle\sum_{\gamma\in\Gamma}(\gamma^{-1}F\pi(h^2)\gamma - \gamma^{-1}\pi(h)F\pi(h)\gamma)\\
&= \textstyle\sum_{\gamma\in\Gamma}\gamma^{-1}(F\pi(h) - \pi(h)F)\pi(h)\gamma\\
&= \textstyle\sum_{\gamma\in\Gamma}\gamma^{-1}(\pi(f)F\pi(h) - \pi(f)\pi(h)F)\pi(h)\pi(f)\gamma\\
&= \textstyle\sum_{\gamma\in\Gamma}\gamma^{-1}\pi(f)[F,\pi(h)]\pi(h)\pi(f)\gamma = \mathcal{A}_f([F,\pi(h)]\pi(h)).
\end{aligned}
$$

In other words $\mathcal{F} - \mathcal{A}_h(F) = \mathcal{A}_f([F,\pi(h)]\pi(h))$; since $[F,\pi(h)]\pi(h)$ is a compact operator by assumption, Lemma 2.12(2) applies to give the result. □

Notice that the above proof really identifies two Kasparov elements, hence gives an equivalent definition for the analytical assembly map:

Corollary 2.16. *Let X be a Γ-compact subset of $\underline{E}\Gamma$; for $x \in KK_i^\Gamma(C_0(X),\mathbb{C})$, one has:*

$$\tilde{\mu}_i^\Gamma(x) = [\mathcal{L}_X] \otimes_{C_0(X)\rtimes\Gamma} j_\Gamma(x).$$

3. Naturality of the Assembly Map

3.1. The case of monomorphisms

Let $\alpha : \Gamma_1 \to \Gamma_2$ be a group monomorphism. We first describe how the Baum–Connes geometric group behaves under α_*. Identifying Γ_1 with $\alpha(\Gamma_1)$, we may assume that Γ_1 is a subgroup of Γ_2 and that α denotes inclusion.

Let X be a Γ_1-compact subset of $\underline{E}\Gamma_1$, and let $x = (\mathcal{H},\pi,F)$ be an element of $KK_i^{\Gamma_1}(C_0(X),\mathbb{C})$, where F is Γ_1-equivariant and properly supported. Our first aim is to describe $\alpha_*(x) \in RK_i^{\Gamma_2}(\underline{E}\Gamma_2)$. Set

$$\tilde{X} = \Gamma_2 \times_{\Gamma_1} X,$$

the quotient of $\Gamma_2 \times X$ by the equivalence relation

$$(\gamma_2\gamma_1, x) \sim (\gamma_2, \gamma_1 x)$$

($\gamma_1 \in \Gamma_1, \gamma_2 \in \Gamma_2, x \in X$); the Γ_2-space $\tilde{X}$ is proper and Γ_2-compact. We denote by $[\gamma_2, x]$ the equivalence class of the pair (γ_2, x). Consider now the Hilbert space

$$\tilde{\mathcal{H}} = \{\xi : \Gamma_2 \to \mathcal{H} : \xi(\gamma_2\gamma_1) = \gamma_1^{-1}(\xi(\gamma_2)) \text{ for every } \gamma_1 \in \Gamma_1,\ \gamma_2 \in \Gamma_2,$$

$$\textit{and} \sum_{\dot{\gamma}\in\Gamma_2/\Gamma_1} \|\xi(\dot{\gamma})\|^2 < \infty\},$$

with Γ_2-action by left translations (this is nothing but the representation induced from Γ_1 to Γ_2).

Notice now that, for $\gamma_2 \in \Gamma_2$, the map

$$\iota_{\gamma_2} : X \to \tilde{X} : x \mapsto [\gamma_2, x]$$

is proper and injective; so for $f \in C_0(\tilde{X})$, the function $f \circ \iota_{\gamma_2}$ belongs to $C_0(X)$. We then define an operator $\tilde{\pi}(f)$ on $\tilde{\mathcal{H}}$ by

$$(\tilde{\pi}(f)\xi)(\gamma_2) = \pi(f \circ \iota_{\gamma_2})(\xi(\gamma_2))$$

($\xi \in \tilde{\mathcal{H}}, \gamma_2 \in \Gamma_2$). It is easy to check that $f \to \tilde{\pi}(f)$ is a Γ_2-covariant representation of $C_0(\tilde{X})$ on $\tilde{\mathcal{H}}$. Define now a Γ_2-equivariant operator $\tilde{F}$ on $\tilde{\mathcal{H}}$ by

$$(\tilde{F}\xi)(\gamma_2) = F(\xi(\gamma_2)).$$

Lemma 3.1. *The triple $(\tilde{\mathcal{H}}, \tilde{\pi}, \tilde{F})$ defines an element $\tilde{x} \in KK_i^{\Gamma_2}(C_0(\tilde{X}), \mathbb{C})$, with $\tilde{F}$ properly supported.*

Proof. We may consider $\tilde{\mathcal{H}}$ as the space of ℓ^2-sections of the Hilbert space bundle $\Gamma_2 \times_{\Gamma_1} \mathcal{H}$ over Γ_2/Γ_1; the choice of a transversal for Γ_2/Γ_1 allows us to trivialize this bundle and to identify (in a *non*-Γ_2-equivariant way!) $\tilde{\mathcal{H}}$ with $\ell^2(\Gamma_2/\Gamma_1) \bigotimes \mathcal{H}$; similarly, the same choice of a transversal identifies topologically $\tilde{X}$ with $(\Gamma_2/\Gamma_1) \times X$; under these identifications, $\tilde{F}$ is realized as $1 \bigotimes F$ and, for $f_1 \in C_0(\Gamma_2/\Gamma_1)$, $f_2 \in C_0(X)$, the operator $\tilde{\pi}(f_1 \bigotimes f_2)$ is realized as $M_{f_1} \bigotimes \pi(f_2)$, where M_{f_1} is the operator of multiplication by f_1 on $\ell^2(\Gamma_2/\Gamma_1)$. Since M_{f_1} is a compact operator, it is clear from this realization that $[\tilde{\pi}(f), \tilde{F}]$ and $\tilde{\pi}(f)(\tilde{F}^2 - 1)$ are compact operators for every $f \in C_0(\tilde{Y})$. It is also clear in this realization that $\tilde{F}$ is properly supported. □

Since $\tilde{X}$ is a proper Γ_2-space, by the universal property of $\underline{E}\Gamma_2$, there exists a Γ_2-equivariant continuous map $\phi : \tilde{X} \to \underline{E}\Gamma_2$. The following lemma is perfectly general.

Lemma 3.2. *Let $\tilde{X}$ be a proper, Γ_2-compact Γ_2-space. Let $\phi : \tilde{X} \to \underline{E}\Gamma_2$ be a continuous, Γ_2-equivariant map. Then ϕ is proper.*

Proof. Let C be a compact subset of $\underline{E}\Gamma_2$: we have to show that $\phi^{-1}(C)$ is compact in $\tilde{X}$. Since $\tilde{X}$ is Γ_2-compact, there exists a compact subset K such that $\tilde{X} = \Gamma_2 \cdot K$. By properness of the Γ_2-action on $\underline{E}\Gamma_2$, the set

$$F = \{\gamma \in \Gamma_2 : C \cap \gamma\phi(K) \neq \emptyset\}$$

is finite. Fix $x \in \phi^{-1}(C)$; let $\gamma \in \Gamma_2$ be such that $\gamma^{-1}x \in K$; then

$$\phi(\gamma^{-1}x) = \gamma^{-1}\phi(x) \in \phi(K) \cap \gamma^{-1}C.$$

This implies $\gamma \in F$, i.e. $x \in F \cdot K$. We have proved $\phi^{-1}(C) \subset F \cdot K$, which shows that $\phi^{-1}(C)$ is compact. □

Set $Y = \phi(\tilde{X})$. Since the map $\phi : \tilde{X} \to Y$ is continuous, proper, Γ_2-equivariant, by functoriality in equivariant K-homology we have $\phi_*(\tilde{x}) = (\tilde{\mathcal{H}}, \tilde{\pi} \circ \phi^*, \tilde{F}) \in KK_i(C_0(Y), \mathbb{C})$, and we set

$$\phi_*(\tilde{x}) = \alpha_*(x).$$

Remark 3.3. By restricting the Γ_2-action on $\underline{E}\Gamma_2$ to Γ_1, we get a universal proper Γ_1-space (see [BCH94], (1.9)). So we could take $\underline{E}\Gamma_1 = \underline{E}\Gamma_2$; in this case, we could take for Y the Γ_2-saturation of X in $\underline{E}\Gamma_2$, and for ϕ the map:

$$\phi : \tilde{X} \to Y : [\gamma_2, x] \mapsto \gamma_2 x.$$

However, we will not assume $\underline{E}\Gamma_1 = \underline{E}\Gamma_2$ in order to get more flexibility (e.g. in Section 4, we deal with the case $\Gamma_1 = \mathbb{Z}$, and we prefer to take $\underline{E}\Gamma_1 = \mathbb{R}$ rather than $\underline{E}\Gamma_1 = \underline{E}\Gamma_2$).

To describe $\tilde{\mu}_i^{\Gamma_2}(\alpha_*(x))$, we have to take the completion $\tilde{\mathcal{E}}$ of $\tilde{\pi}(\phi^*(C_c(Y)))\tilde{\mathcal{H}}$ with respect to the $\mathbb{C}\Gamma_2$-valued scalar product

$$\langle \xi_1 | \xi_2 \rangle (\gamma_2) = \langle \xi_1 | \gamma_2 \xi_2 \rangle$$

($\xi_1, \xi_2 \in \tilde{\pi}(\phi^*(C_c(Y)))\mathcal{H}$, $\gamma_2 \in \Gamma_2$), and with the operator $\tilde{\mathcal{F}}$ that extends $\tilde{F}$ continuously (see Lemma 2.8).

On the other hand, to describe $\alpha_*(\tilde{\mu}_i^{\Gamma_1}(x))$, we first consider (as in Lemma 2.7) the completion $\mathcal{E}$ of $\pi(C_c(X))\mathcal{H}$ with respect to the $\mathbb{C}\Gamma_1$-valued scalar product

$$\langle \eta_1 | \eta_2 \rangle (\gamma_1) = \langle \eta_1 | \gamma_1 \eta_2 \rangle$$

($\eta_1, \eta_2 \in \mathcal{H}$, $\gamma_1 \in \Gamma_1$). The element $\alpha_*(\tilde{\mu}_i^{\Gamma_1}(x))$ is then described by the pair $(\mathcal{E} \bigotimes_{C^*\Gamma_1} C^*\Gamma_2, \mathcal{F} \bigotimes 1)$, where $C^*\Gamma_2$ is viewed as a left module over $C^*\Gamma_1$, and as a right module over itself. The $C^*\Gamma_2$-valued scalar product on $\mathcal{E} \bigotimes_{C^*\Gamma_1} C^*\Gamma_2$ is given by

$$\langle \eta_1 \otimes b_1 | \eta_2 \otimes b_2 \rangle_{C^*\Gamma_2} = b_1^* \langle \eta_1 | \eta_2 \rangle_{C^*\Gamma_1} b_2$$

($\eta_1, \eta_2 \in \mathcal{E}$, $b_1, b_2 \in C^*\Gamma_2$; see [Lan95], 4.5).

Proof of Theorem 1.1, *case of a monomorphism.* We have to show $\alpha_*(\tilde{\mu}_i^{\Gamma_1}(x)) = \tilde{\mu}_i^{\Gamma_2}(\alpha_*(x))$ in $KK_i(\mathbb{C}, C^*\Gamma_2) = K_i(C^*\Gamma_2)$. Following a similar construction due to Rieffel ([Rie74], p. 228), we define a map

$$\Psi : \begin{cases} \pi(C_c(X))\mathcal{H} \bigotimes_{\mathbb{C}\Gamma_1} \mathbb{C}\Gamma_2 & \to & \tilde{\mathcal{H}} \\ \eta \otimes b & \mapsto & (\gamma_2 \mapsto \sum_{\gamma_1 \in \Gamma_1} b(\gamma_1^{-1}\gamma_2^{-1})\gamma_1 \eta \end{cases}$$

($\eta \in \pi(C_c(X))\mathcal{H}$, $b \in \mathbb{C}\Gamma_2$, $\gamma_2 \in \Gamma_2$). It follows from Theorem 5.12 in [Rie74] that Ψ is well-defined, i.e., for $\gamma_1 \in \Gamma_1$:

$$\Psi(\eta\gamma_1 \otimes b) = \Psi(\eta \otimes \gamma_1 b).$$

It is readily checked that Ψ is a $\mathbb{C}\Gamma_2$-module map. Now, for $\gamma, \gamma_2 \in \Gamma_2$, one has:

$$\Psi(\eta \otimes \gamma)(\gamma_2) = \begin{cases} \gamma_2^{-1}\gamma^{-1}\eta & if\ \gamma.\gamma_2 \in \Gamma_1 \\ 0 & otherwise \end{cases}$$

It also follows from this that the range of Ψ is contained in $\tilde{\pi}(\phi^*(C_c(Y)))\tilde{\mathcal{H}}$; to see it, suppose that $\eta = \pi(f)\xi$ (with $f \in C_c(X)$, $\xi \in \mathcal{H}$), and choose $g \in C_c(Y)$ such that $g = 1$ on $\gamma^{-1}(\phi(\iota_1(supp\, f)))$. Then

$$\Psi(\eta \otimes \gamma) = \tilde{\pi}(\phi^* g)\Psi(\eta \otimes \gamma) \in \tilde{\pi}(\phi^*(C_c(Y)))\tilde{\mathcal{H}}$$

Let $\{s_i\}_{i \in I}$ be a transversal for Γ_2/Γ_1. The inverse map

$$\Psi^{-1} : \tilde{\pi}(\phi^*(C_c(Y)))\tilde{\mathcal{H}} \to \pi(C_c(X))\mathcal{H} \otimes_{\mathbb{C}\Gamma_1} \mathbb{C}\Gamma_2$$

is given by:

$$\Psi^{-1}(\tilde{\xi}) = \sum_{i \in I} \tilde{\xi}(s_i) \otimes s_i^{-1} \tag{3.1}$$

($\tilde{\xi} \in \tilde{\pi}(\phi^*(C_c(Y)))\tilde{\mathcal{H}}$); using the fact that there are finitely many s_i's such that $\iota_{s_i}(X)$ meets a given compact subset of $\tilde{X}$, one sees that the sum in (3.1) is actually a finite sum. Next, for $\eta, \eta' \in \pi(C_c(X))\mathcal{H}$ and $\gamma, \gamma', \gamma_2 \in \Gamma_2$, we compute:

$$\begin{aligned}
\langle \Psi(\eta \otimes \gamma) | \Psi(\eta' \otimes \gamma') \rangle(\gamma_2) &= \langle \Psi(\eta \otimes \gamma) | \gamma_2 \Psi(\eta' \otimes \gamma') \rangle \\
&= \sum_{\dot{\sigma} \in \Gamma_2/\Gamma_1} \langle \Psi(\eta \otimes \gamma) | \Psi(\eta' \otimes \gamma')(\gamma_2^{-1}\sigma) \rangle \\
&= \sum_{\substack{\dot{\sigma} \in \Gamma_2/\Gamma_1 \\ \gamma\sigma \in \Gamma_1 \\ \gamma'\gamma_2^{-1}\sigma \in \Gamma_1}} \langle \sigma^{-1}\gamma\eta | \sigma^{-1}\gamma_2\gamma'\eta' \rangle \\
&= \begin{cases} \langle \eta | \gamma\gamma_2\gamma'^{-1}\eta' \rangle & if\ \gamma\gamma_2\gamma'^{-1} \in \Gamma_1 \\ 0 & otherwise \end{cases} \\
&= \langle \eta | \eta' \rangle_{C^*\Gamma_1}(\gamma\gamma_2\gamma'^{-1}) \\
&= (\gamma^{-1}\langle \eta | \eta' \rangle_{C^*\Gamma_1}\gamma')(\gamma_2) \\
&= \langle \eta \otimes \gamma | \eta' \otimes \gamma' \rangle_{C^*\Gamma_2}(\gamma_2).
\end{aligned}$$

This means that Ψ extends to a unitary isomorphism of $C^*\Gamma_2$-modules between $\mathcal{E} \otimes_{C^*\Gamma_1} C^*\Gamma_2$ and $\tilde{\mathcal{E}}$. Moreover, an easy computation gives

$$\Psi(\mathcal{F} \bigotimes 1)\Psi^{-1} = \tilde{\mathcal{F}},$$

which concludes the proof. □

3.2. The case of epimorphisms

Let $\alpha : \Gamma_1 \to \Gamma_2$ be a group epimorphism; set $N = \ker \alpha$. By identifying Γ_1/N with Γ_2, we may assume that α is the quotient-map. We also denote by α the induced algebra homomorphisms $\mathbb{C}\Gamma_1 \twoheadrightarrow \mathbb{C}\Gamma_2$ and $C^*\Gamma_1 \twoheadrightarrow C^*\Gamma_2$.

We first describe how the left-hand side of the assembly map behaves under α_*; we were inspired by Kasparov's descent homomorphism in Theorem 3.4 of [Kas88]. Let then X be a Γ_1-compact subset of $\underline{E}\Gamma_1$, and $x = (\mathcal{H}, \pi, F)$ be an element of $KK_i^{\Gamma_1}(C_0(X), \mathbb{C})$, with F properly supported and Γ_1-equivariant. Set $\tilde{X} = N \backslash X$: this is a proper, Γ_2-compact Γ_2-space.

Consider on $\pi(C_c(X))\mathcal{H}$ the scalar product

$$\ll \xi|\eta \gg = \sum_{n\in N} \langle \xi|n\eta\rangle$$

($\xi, \eta \in \pi(C_c(X))\mathcal{H}$); it is non-negative, since by Lemma 2.7 the function $n \mapsto \langle \xi|n\xi\rangle$ defines a positive element in C^*N; so by applying the trivial representation we get $\ll \xi|\xi \gg \geq 0$. Let $\tilde{\mathcal{H}}$ be the separation-completion of $\pi(C_c(X))\mathcal{H}$ for this scalar product. The natural action of Γ_1 on $\pi(C_c(X))\mathcal{H}$ is isometric (because N is normal in Γ_1); hence it extends to a unitary representation of Γ_1 on $\tilde{\mathcal{H}}$ which is trivial on N, hence factors through a unitary representation of Γ_2. Let T be a bounded operator on $\mathcal{H}$, preserving $\pi(C_c(X))\mathcal{H}$ and N-equivariant; it follows from Lemma 2.8 that, for some $K > 0$ and every $\xi \in \pi(C_c(X))\mathcal{H}$, the element

$$K\langle \xi|\xi\rangle(\cdot) - \langle T\xi|T\xi\rangle(\cdot)$$

is a positive element in C^*N. Summing over N, this implies

$$\ll T\xi|T\xi \gg \leq K \ll \xi|\xi \gg,$$

so that T extends to a bounded operator $\tilde{T}$ on $\tilde{\mathcal{H}}$. In particular, F provides a Γ_2-equivariant operator $\tilde{F}$ on $\tilde{\mathcal{H}}$.

Any function on $\tilde{X}$ can be lifted to an N-invariant function on X. Viewing in this way $C_0(\tilde{X})$ as an algebra of multipliers of $C_0(X)$, and extending the representation π to multipliers, as in [Ped79] 3.12.10, we get an algebra of operators on $\mathcal{H}$ that preserve $\pi(C_c(X))\mathcal{H}$ and commute with N, so that the preceding observation provides a Γ_2-covariant representation $\tilde{\pi}$ of $C_0(\tilde{X})$ on $\tilde{\mathcal{H}}$.

Lemma 3.4. *The triple $(\tilde{\mathcal{H}}, \tilde{\pi}, \tilde{F})$ defines an element $\tilde{x} \in KK_i^{\Gamma_2}(C_0(\tilde{X}), \mathbb{C})$, with $\tilde{F}$ properly supported.*

Proof. We consider the C^*N-module $\mathcal{E}$ obtained by completing $\pi(C_c(X))\mathcal{H}$ with respect to the scalar product given by equation (2.4) (with Γ replaced by N); it is clear from the definitions that

$$\tilde{\mathcal{H}} = \mathcal{E} \otimes_{C^*N} \mathbb{C},$$

where $\mathbb{C}$ is a left C^*N-module via the augmentation map (i.e. the character coming from the trivial representation of N). For $T \in \mathcal{L}(H)$ preserving $\pi(C_c(X))\mathcal{H}$ and N-equivariant, one has, using the notation from Lemma 2.8:

$$\tilde{T} = T \otimes 1.$$

We have to prove that, for every $\tilde{h} \in C_0(\tilde{X})$, the operators $[\tilde{\pi}(\tilde{h}), \tilde{F}]$ and $\tilde{\pi}(\tilde{h})(\tilde{F}^2 - 1)$ are compact. By linearity and density, we may assume that $\tilde{h}$ belongs to $C_c(\tilde{X})$ and that $\tilde{h}$ is real-valued. Find $h \in C_c(X)$, real-valued, such that

$$\tilde{h} = \sum_{n\in N} n(h). \tag{3.2}$$

As in the proof of Proposition 2.9, let $f \in C_c(X)$ be a plateau function such that $f.h = h$ and $\pi(f)F\pi(h) = F\pi(h)$. Then

$$\begin{aligned}[\pi(\tilde{h}), F] &= \sum_{n\in N} n[\pi(h), F]n^{-1} \\ &= \sum_{n\in N} n\pi(f)[\pi(h), F]\pi(f)n^{-1} \\ &= A_f([\pi(h), F])\end{aligned}$$

(notation as in Lemma 2.12). It is a consequence of Lemma 2.12(2) that the extension $\mathcal{A}_f([\pi(h), F])$ of $A_f([\pi(h), F])$ to $\mathcal{E}$, is a compact operator on $\mathcal{E}$. Since

$$[\tilde{\pi}(\tilde{h}), \tilde{F}] = \mathcal{A}_f([\pi(h), F]) \otimes 1,$$

it follows from 4.7 in [Lan95], that $[\tilde{\pi}(\tilde{h}), \tilde{F}]$ is compact on $\mathcal{H}$. A similar argument works for compactness of $\tilde{\pi}(\tilde{h})(\tilde{F}^2 - 1)$.
It remains to prove that $\tilde{F}$ is properly supported. Thus, fix $\tilde{h} \in C_c(\tilde{X})$, and choose $h \in C_c(X)$ as in (3.2). Since F is properly supported, one finds $g \in C_c(X)$ such that $\pi(g)F\pi(h) = F\pi(h)$ and $g.h = h$. As above, one computes:

$$F\pi(\tilde{h}) = A_g(F\pi(h)).$$

Let $\tilde{K}$ be the image of *supp* g in $\tilde{X}$; let $\tilde{f} \in C_c(\tilde{X})$ be equal to 1 on $\tilde{K}$, so that $\tilde{f}.g = g$ (where $\tilde{f}$ is now viewed as a multiplier of $C_0(X)$). Since $\pi(\tilde{f})$ is N-equivariant, one has

$$\pi(\tilde{f})A_g(F\pi(h)) = A_g(F\pi(h)),$$

i.e.

$$\tilde{\pi}(\tilde{f})\tilde{F}\tilde{\pi}(\tilde{h}) = \tilde{F}\tilde{\pi}(\tilde{h}).$$

This concludes the proof of the lemma. □

Since $\tilde{X}$ is a proper, Γ_2-compact Γ_2-space, there exists a Γ_2-equivariant continuous map $\phi : \tilde{X} \to \underline{E}\Gamma_2$, which is unique up to Γ_2-equivariant homotopy. By Lemma 3.2, this map ϕ is proper. Set $Y = \phi(\tilde{X})$; applying functoriality in equivariant K-homology we get $\phi_*(\tilde{x}) = (\tilde{\mathcal{H}}, \tilde{\pi} \circ \phi^*, \tilde{F}) \in KK_i^{\Gamma_2}(C_0(Y), \mathbb{C})$, and we set

$$\phi_*(\tilde{x}) = \alpha_*(x).$$

Remark 3.5. The assumption of Γ_1-compactness was used in the proof of Lemma 3.4 only to ensure that the space X is locally compact. So, if X is a locally compact, proper Γ_1-space, what Lemma 3.4 actually does is constructing a homomorphism

$$\alpha_* : KK_i^{\Gamma_1}(C_0(X), \mathbb{C}) \to KK_i^{\Gamma_2}(C_0(N\backslash X), \mathbb{C}).$$

We record this for future reference.

Example 3.6. Let $\alpha : \Gamma \to \{1\}$ be the only homomorphism to the trivial group; then $\alpha : C^*\Gamma \to \mathbb{C}$ is the augmentation map, i.e. the character of the trivial representation. It follows from the proof of Lemma 3.4 that the "geometric" map $\alpha_{*,g} : RK_0^\Gamma(\underline{E}\Gamma) \to K_0(pt) = \mathbb{Z}$ is actually given by

$$\alpha_{*,g} = \alpha_{*,a} \circ \tilde{\mu}_0^\Gamma,$$

where $\alpha_{*,a} : K_0(C^*\Gamma) \to K_0(\mathbb{C}) = \mathbb{Z}$ is the "analytical" map. This means that, in this case, Theorem 1.1 is essentially built in the definition of $\alpha_{*,g}$. *Assume that the Baum–Connes conjecture is true for the group* Γ. Then $\lambda_\Gamma : K_0(C^*\Gamma) \to K_0(C_r^*\Gamma)$ is onto, with a canonical splitting given by $\tilde{\mu}_0^\Gamma \circ (\mu_0^\Gamma)^{-1}$; loosely speaking, there is a canonical copy of $K_0(C_r^*\Gamma)$ inside $K_0(C^*\Gamma)$; this means that *any* representation of Γ, and in particular the trivial representation α, is defined on $K_0(C_r^*\Gamma)$. It seems to be an interesting *question* to define, without appealing to the Baum–Connes conjecture, a map

$$\alpha_? : K_0(C_r^*\Gamma) \to \mathbb{C}$$

such that, on the image of $\tilde{\mu}_0^\Gamma$, one has:

$$\alpha_{*,a} = \alpha_? \circ (\lambda_\Gamma)_*.$$

If Γ is torsion-free, then the answer is given by $\alpha_? = (\tau_\Gamma)_*$, the homomorphism $K_0(C_r^*\Gamma) \to \mathbb{C}$ associated with the canonical trace τ_Γ on $C_r^*\Gamma$. Indeed the equality of the two maps $\alpha_{*,a} \circ \tilde{\mu}_0^\Gamma$ and $(\tau_\Gamma)_* \circ \mu_0^\Gamma$, from $RK_0^\Gamma(E\Gamma)$ to $\mathbb{Z}$, is proved in Theorem 3.3.1 in [Pie00] (see also Theorem 7.14 of [Mis]), and can be seen as a version of Atiyah's L^2-index theorem.

Proof of Theorem 1.1, *case of an epimorphism.* We have to show $\alpha_*(\tilde{\mu}_i^{\Gamma_1}(x)) = \tilde{\mu}_i^{\Gamma_2}(\alpha_*(x))$ in $KK_i(\mathbb{C}, C^*\Gamma_2) = K_i(C^*\Gamma_2)$.

The Hilbert $C^*\Gamma_2$-module underlying $\alpha_*(\tilde{\mu}_i^{\Gamma_1}(x))$ is (as in Section 3.1) the tensor product $\mathcal{E} \otimes_{C^*\Gamma_1} C^*\Gamma_2$, where $C^*\Gamma_2$ is viewed as a left module over $C^*\Gamma_1$ via α, and as a right module over itself. The $C^*\Gamma_2$-valued scalar product on $\mathcal{E} \otimes_{C^*\Gamma_1} C^*\Gamma_2$ is given by

$$\langle \eta_1 \otimes b_1 | \eta_2 \otimes b_2 \rangle_{C^*\Gamma_2} = b_1^* \alpha(\langle \eta_1 | \eta_2 \rangle_{C^*\Gamma_1}) b_2$$

($\eta_1, \eta_2 \in \mathcal{E}$, $b_1, b_2 \in C^*\Gamma_2$; see 4.5 in [Lan95]). The operator giving the K-theory element is $\mathcal{F} \otimes 1$.

On the other hand the $C^*\Gamma_2$-module underlying $\tilde{\mu}_i^{\Gamma_2}(\alpha_*(x))$ is defined as the completion $\tilde{\mathcal{E}}$ of $\tilde{\pi}(\phi^*(C_c(Y)))\tilde{\mathcal{H}}$ with respect to the $\mathbb{C}\Gamma_2$-scalar product

$$\ll \xi_1 | \xi_2 \gg (\gamma_2) = \ll \xi_1 | \gamma_2 \xi_2 \gg$$

($\xi_1, \xi_2 \in \tilde{\pi}(\phi^*(C_c(Y)))\tilde{\mathcal{H}}$, $\gamma_2 \in \Gamma_2$). The operator $\tilde{\mathcal{F}}$ giving the K-theory element is the continuous extension of $\tilde{F}$ given by Lemma 2.8.

For $\eta \in \pi(C_c(X))\mathcal{H}$, we denote by $\tilde{\eta}$ the image of η in $\tilde{\mathcal{H}}$. We notice that $\tilde{\eta}$ really belongs to $\tilde{\pi}(\phi^*(C_c(Y)))\tilde{\mathcal{H}}$; indeed, for $\eta = \pi(f)\xi$, let K be the image of *supp* f in $\tilde{X}$, and let $g \in C_c(Y)$ be equal to 1 on $\phi(K)$. Then $\phi^*(g).f = f$ and

$$\tilde{\pi}(\phi^* g)\widetilde{\pi(f)\xi} = \widetilde{\pi(f)\xi} = \tilde{\eta}.$$

We define a map

$$\Psi : \begin{cases} \pi(C_c(X)\mathcal{H} \otimes_{\mathbb{C}\Gamma_1} \mathbb{C}\Gamma_2 & \to & \tilde{\pi}(\phi^*(C_c(Y)))\tilde{\mathcal{H}} \\ \eta \otimes \gamma_2 & \mapsto & \gamma_2^{-1}\tilde{\eta} \end{cases}$$

($\eta \in \pi(C_c(X)\mathcal{H}$, $\gamma_2 \in \Gamma_2$). Next, for $\gamma, \gamma', \gamma_2 \in \Gamma_2$, let $\gamma_1 \in \Gamma_1$ be any group element such that $\alpha(\gamma_1) = \gamma\gamma_2\gamma'^{-1}$. Then, for $\eta, \eta' \in \pi(C_c(X))\mathcal{H}$ we compute:

$$\begin{aligned} \langle \eta \otimes \gamma | \eta' \otimes \gamma' \rangle(\gamma_2) &= (\gamma^{-1}\alpha(\langle \eta|\eta' \rangle_{C^*\Gamma_1})\gamma')(\gamma_2) \\ &= \alpha(\langle \eta|\eta' \rangle_{C^*\Gamma_1})(\gamma\gamma_2\gamma'^{-1}) \\ &= \textstyle\sum_{n \in N} \langle \eta|\eta' \rangle_{C^*\Gamma_1}(n\gamma_1) \\ &= \textstyle\sum_{n \in N} \langle \eta|n\gamma_1\eta' \rangle \\ &= \ll \tilde{\eta}|\gamma\gamma_2\gamma'^{-1}\tilde{\eta'} \gg \\ &= \ll \gamma^{-1}\tilde{\eta}|\gamma'^{-1}\tilde{\eta'} \gg (\gamma_2) \\ &= \ll \Psi(\eta \otimes \gamma)|\Psi(\eta' \otimes \gamma') \gg (\gamma_2) \end{aligned}$$

This means that Ψ extends to a unitary isomorphism of $C^*\Gamma_2$-modules between $\mathcal{E} \otimes_{C^*\Gamma_1} C^*\Gamma_2$ and $\tilde{\mathcal{E}}$. Notice that, for $f \in C_c(X)$, $g \in C_c(Y)$, $\xi \in \mathcal{H}$, one has:

$$\tilde{\pi}(\phi^* g)\widetilde{\pi(f)\xi} = \widetilde{\pi(\phi^* g.f)\xi} = \Psi(\pi(\phi^* g.f)\xi \otimes 1).$$

This shows that Ψ has dense image, so that Ψ is onto. Finally, an easy computation gives

$$\Psi(\mathcal{F} \otimes 1)\Psi^{-1} = \tilde{\mathcal{F}},$$

which concludes the proof of Theorem 1.1. □

3.3. An application to free actions

Let X be a locally compact, proper Γ-space; let α be the homomorphism from Γ to the trivial group. By Remark 3.5, there is a homomorphism

$$\alpha_* : KK_i^\Gamma(C_0(X), \mathbb{C}) \to KK_i(C_0(\Gamma\backslash X), \mathbb{C}).$$

Corollary 3.7. *If Γ acts properly and freely on X, then α_* is an isomorphism.*

It is of course well-known that, in the case of a free action:

$$KK_i^\Gamma(C_0(X), \mathbb{C}) \simeq KK_i(C_0(\Gamma\backslash X), \mathbb{C})$$

(see [Rie82]). This is usually proved by identifying $KK_i^\Gamma(C_0(X), \mathbb{C})$ with $KK_i(C_0(X) \rtimes \Gamma, \mathbb{C})$, and then appealing to freeness of the action to conclude that $C_0(X) \rtimes \Gamma$ is Morita equivalent to $C_0(\Gamma\backslash X)$. We think that the interest of Corollary 1.3 is to provide an explicit and easily described isomorphism. Thanks are due to S. Echterhoff for his help in the following proof.

Proof of Corollary 3.7. We are going to show that α_* coincides with the isomorphism obtained via Morita equivalence, as indicated above. The imprimitivity bimodule realizing the Morita equivalence between $C_0(\Gamma\backslash X)$ and $C_0(X) \rtimes \Gamma$ is a suitable completion $\overline{C_c(X)}$ of $C_c(X)$, with the obvious left action of $C_c(\Gamma\backslash X)$, the right $C_c(X \times \Gamma)$-module structure given by formula (2.6), and the $C_c(X \times \Gamma)$-valued scalar product given by formula (2.7). The bimodule $\overline{C_c(X)}$ defines an element $[\overline{C_c(X)}]$ in the Kasparov group $KK_0(C_0(\Gamma\backslash X), C_0(X) \rtimes \Gamma)$.

Consider the Kasparov elements

$$x = (\mathcal{H}, \pi, F) \in KK_i^{\Gamma}(C_0(X), \mathbb{C})$$

and

$$\alpha_*(x) = (\tilde{\mathcal{H}}, \tilde{\pi}, \tilde{F}) \in KK_i(C_0(\Gamma\backslash X), \mathbb{C}),$$

as in Lemma 3.4. View x as an element of $KK_i(C_0(X) \rtimes \Gamma, \mathbb{C})$. We want to show that

$$[\overline{C_c(X)}] \otimes_{C_0(X)\rtimes\Gamma} x \;=\; \alpha_*(x). \tag{3.3}$$

For $f \in C_c(X)$ and $\xi \in \mathcal{H}$, denote by $[\pi(f)\xi]$ the image of $\pi(f)\xi$ in $\tilde{\mathcal{H}}$. An easy check shows that the map

$$C_c(X) \otimes \mathcal{H} \to \tilde{\mathcal{H}} : \; f \otimes \xi \mapsto [\pi(f)\xi]$$

extends to a unitary isomorphism between the Hilbert spaces $\overline{C_c(X)} \otimes_{C_0(X)\rtimes\Gamma} \mathcal{H}$ and $\tilde{\mathcal{H}}$, which moreover intertwines the representations of $C_0(\Gamma\backslash X)$.

To show that the operator $\tilde{F}$ realizes the Kasparov product, we use the connection formalism of Connes and Skandalis [CS84]. For $f \in C_c(X)$, consider the map

$$\theta_f : \mathcal{H} \to \tilde{\mathcal{H}} : \xi \mapsto [\pi(f)\xi].$$

We have to show that $\tilde{F}$ is an F-connection, i.e. that the operator $\theta_f F - \tilde{F}\theta_f$ is a compact operator from $\mathcal{H}$ to $\tilde{\mathcal{H}}$. Set $T = [\pi(f), F]$, a compact operator on $\mathcal{H}$. Notice that, for $\xi \in \pi(C_c(X))\mathcal{H}$, one has

$$(\theta_f F - \tilde{F}\theta_f)\xi \;=\; [T\xi]. \tag{3.4}$$

Since F is properly supported, there exists $g \in C_c(X)$ such that $\pi(g)T\pi(g) = T$.

We claim that the map

$$\overline{\pi(g)\mathcal{H}} \to \tilde{\mathcal{H}} : \pi(g)\eta \mapsto [\pi(g)\eta]$$

is bounded. Indeed

$$\ll \pi(g)\eta | \pi(g)\eta \gg \, = \sum_{\gamma\in\Gamma} \langle \gamma\pi(g)\eta | \pi(g)\eta \rangle .$$

But the summation in the right hand side is over the finite set

$$F = \{\gamma \in \Gamma : \gamma(supp\, g) \cap supp\, g \neq \emptyset\},$$

so that, by the Cauchy–Schwarz inequality:

$$\ll \pi(g)\eta | \pi(g)\eta \gg \, \leq card\, F \cdot \|\pi(g)\eta\|^2,$$

which proves the claim.

Consider now the product of the following three operators:

$$\begin{array}{ll} \mathcal{H} \to \overline{\pi(g)\mathcal{H}}: & \xi \mapsto \pi(g)\xi; \\ \overline{\pi(g)\mathcal{H}} \to \overline{\pi(g)\mathcal{H}} & \zeta \mapsto T\zeta; \\ \overline{\pi(g)\mathcal{H}} \to \tilde{\mathcal{H}} & \pi(g)\eta \mapsto [\pi(g)\eta]. \end{array}$$

It follows from equation (3.4) that this product is exactly $\theta_f F - \tilde{F}\theta_f$. Since T is compact and the two other operators are bounded, it follows that $\theta_f F - \tilde{F}\theta_f$ is compact: this proves formula (3.3), and hence concludes the proof. □

4. Interlude: The Group of Integers

For the group $\mathbb{Z}$ of integers, we take $\underline{E}\mathbb{Z} = E\mathbb{Z} = \mathbb{R}$ (the real line) and $B\mathbb{Z} = S^1$ (the circle). By Fourier series (see [Ped79], 7.1.6), we identify $C^*\mathbb{Z}$ with $C(S^1)$.

Proposition 4.1. *For $i = 0, 1$, the map*

$$\mu_i^{\mathbb{Z}} : K_i^{\mathbb{Z}}(\mathbb{R}) = K_i(S^1) \to K_i(C^*\mathbb{Z})$$

is an isomorphism.

4.1. Proof, case $i = 0$

This is the easy case. We have $K_0(S^1) = \mathbb{Z}$, generated by the character of $C(S^1)$ given by evaluation at a given base-point (or, dually, generated by the inclusion of the base-point). When lifted to $\mathbb{R}$, this gives exactly the element $\beta_{\{0\}} \in K_0^{\mathbb{Z}}(\mathbb{R})$ described in Example 2.11. On the other hand, $K_0(C(S^1)) = \mathbb{Z}$, generated by the class of the constant function 1 (or, dually, generated by the trivial one-dimensional vector bundle over S^1). The result then follows from Example 2.11 above. □

4.2. Proof, case $i = 1$

Here again, both groups $K_1(S^1)$ and $K_1(C(S^1))$ are isomorphic to $\mathbb{Z}$. To describe the generator of $K_1(S^1)$, we identify S^1 with $\mathbb{R}/\mathbb{Z}$ and consider the Hilbert space $L^2(S^1)$ with the trigonometric basis $(\exp(2\pi i n\theta))_{n\in\mathbb{Z}}$. Consider the operator F, diagonal in that basis, given by

$$F = diag(sign(n))_{n\in\mathbb{Z}}.$$

Let M be the representation of $C(S^1)$ by pointwise multiplication on $L^2(S^1)$. Then the triple $(L^2(S^1), M, F)$ defines the "Toeplitz" generator of $K_1(S^1) = KK_1(C(S^1), \mathbb{C})$.

To proceed, it will be convenient to work in the context of unbounded Kasparov elements (in the sense of Baaj–Julg [BJ83]). The unbounded picture of $(L^2(S^1), M, F)$ is $(L^2(S^1), M, D)$, where

$$D = \frac{1}{i} \cdot \frac{d}{d\theta};$$

indeed the phase of the operator D is just F. To say that $(L^2(S^1), M, D)$ is an *unbounded Kasparov module* means that:

1. D is a densely defined, self-adjoint operator;
2. $M(f)(1+D^2)^{-1}$ is a compact operator for every $f \in C(S^1)$;
3. $[M(f), D]$ is a bounded operator for every f in a dense subalgebra of $C(S^1)$ (here $C^\infty(S^1)$).

Working with D has the advantage of being independant of the choice of any particular basis.

Via the isomorphism $K_1(S^1) \simeq K_1^{\mathbb{Z}}(\mathbb{R})$, the triple $(L^2(S^1), M, D)$ goes to the triple $(L^2(\mathbb{R}), \tilde{M}, \tilde{D})$, where $\tilde{M}$ is the $\mathbb{Z}$-covariant representation of $C_0(\mathbb{R})$ by pointwise multiplication on $L^2(\mathbb{R})$, and

$$\tilde{D} = \frac{1}{i} \cdot \frac{d}{dt}.$$

As a domain for $\tilde{D}$, we take the Schwartz space $\mathcal{S}(\mathbb{R})$, i.e the space of C^∞-functions on $\mathbb{R}$ which have rapid decay, together with all their derivatives.

We pause at this point to notice that, since the Fourier transform of the operator $\tilde{D}$ is the operator $\tilde{E}$ of multiplication by the dual variable λ (up to a factor 2π), the most natural bounded version of $\tilde{D}$ would be H, the convolution operator by the function whose Fourier transform is the sign function, which is nothing but the Hilbert transform on $L^2(\mathbb{R})$. One difficulty here is that H is *not* properly supported! (This can be seen as a good reason to consider the unbounded picture for Kasparov elements...)

To proceed, we shall appeal in a systematic way to the Fourier transform $\xi \mapsto \hat{\xi}$, where for $\xi \in \mathcal{S}(\mathbb{R})$ and $\lambda \in \mathbb{R}$:

$$\hat{\xi}(\lambda) = \int_{\mathbb{R}} \xi(t) \exp(-2\pi i \lambda t)\, dt.$$

Fourier transforming the Kasparov triple $(L^2(\mathbb{R}), \tilde{M}, \tilde{D})$ yields the Kasparov triple $(L^2(\mathbb{R}), \Lambda, \tilde{E})$ where Λ is the representation of $C_0(\mathbb{R})$ by convolution by Fourier transforms, and $n \in \mathbb{Z}$ now acts (on the left) by pointwise multiplication by the function $\lambda \mapsto \exp(-2\pi i n \lambda)$. The domain of $\tilde{E}$ is $\mathcal{S}(\mathbb{R})$. The $C^*\mathbb{Z}$-valued scalar product on $\mathcal{S}(\mathbb{R})$ is given by

$$\langle \xi_1 | \xi_2 \rangle(n) \;=\; \langle \xi_1 | n(\xi_2) \rangle \;=\; \int_{\mathbb{R}} \overline{\xi_1(\lambda)} \exp(-2\pi i n \lambda) \xi_2(\lambda)\, d\lambda$$

($\xi_1, \xi_2 \in \mathcal{S}(\mathbb{R})$, $n \in \mathbb{Z}$). Under the identification $C^*\mathbb{Z} \to C(S^1)$ given by Fourier series, i.e.

$$a \mapsto (\theta \mapsto \sum_{n \in \mathbb{Z}} a(n) \exp(2\pi i n \theta))$$

($a \in \mathbb{C}\mathbb{Z}$, $\theta \in S^1$), this becomes a $C(S^1)$-valued scalar product:

$$\langle \xi_1 | \xi_2 \rangle(\theta) = \sum_{n \in \mathbb{Z}} \int_{\mathbb{R}} \overline{\xi_1(\lambda)} \exp(2\pi i n(\theta - \lambda)) \xi_2(\lambda)\, d\lambda.$$

Let $\mathcal{E}$ be the completion of $\mathcal{S}(\mathbb{R})$ for this $C(S^1)$-valued scalar product. Then $\mu_1^{\mathbb{Z}}(L^2(\mathbb{R}), \Lambda, \tilde{E})$ is described by $(\mathcal{E}, \tilde{E}) \in KK_1(\mathbb{C}, C(S^1))$. We now want to describe the Hilbert $C(S^1)$-module $\mathcal{E}$ as a continuous field of Hilbert spaces over S^1.

For that purpose, consider the Hilbert bundle over S^1 induced by the left regular representation of $\mathbb{Z}$; in other words, the total space of this bundle is $\mathbb{R} \times_{\mathbb{Z}} \ell^2(\mathbb{Z})$ and the space of continuous sections is

$$\mathcal{E}' = \{\eta : \mathbb{R} \to \ell^2(\mathbb{Z}),\ \text{continuous}\,,\ \eta(\lambda+1)_n = \eta(\lambda)_{n+1} \text{ for every } n \in \mathbb{Z},\ \lambda \in \mathbb{R}\};$$

a section is determined by its values for $-\frac{1}{2} \leq \lambda < \frac{1}{2}$. Pointwise scalar product turns $\mathcal{E}'$ into a $C(S^1)$-Hilbert module:

$$\langle \eta_1 | \eta_2 \rangle(\theta) = \sum_{n \in \mathbb{Z}} \overline{\eta_1(\tilde{\theta})_n \eta_2(\tilde{\theta})_n}$$

($\eta_1, \eta_2 \in \mathcal{E}'$); here $\tilde{\theta}$ is any real number which lifts $\theta \in S^1$.

Consider the map $\Psi : \mathcal{S}(\mathbb{R}) \to \mathcal{E}'$ defined by

$$((\Psi(\xi))(\lambda))_n = \xi(\lambda + n)$$

($\xi \in \mathcal{S}(\mathbb{R})$, $\lambda \in \mathbb{R}$, $n \in \mathbb{Z}$). Let $n \in \mathbb{Z}$ act on $\mathcal{E}'$ (on the right) by pointwise multiplication by $\theta \mapsto \exp(2\pi i n\theta)$. It is clear that Ψ is a $\mathbb{C}\mathbb{Z}$-module map. Moreover Ψ is isometric with respect to $C(S^1)$-valued scalar products; indeed:

$$\begin{aligned} \langle \Psi(\xi_1) | \Psi(\xi_2) \rangle(\theta) &= \textstyle\sum_{n \in \mathbb{Z}} \overline{\xi_1(\tilde{\theta}+n)} \xi_2(\tilde{\theta}+n) \\ &= \textstyle\sum_{n \in \mathbb{Z}} (\overline{\xi_1}\xi_2)(\tilde{\theta}+n) \\ &= \textstyle\sum_{n \in \mathbb{Z}} \widehat{\overline{\xi_1}\xi_2}(n) \exp(2\pi i n\theta) \\ &= \textstyle\sum_{n \in \mathbb{Z}} \exp(2\pi i n\theta) \int_{\mathbb{R}} (\overline{\xi_1}\xi_2)(\lambda) \exp(-2\pi i n\lambda)\, d\lambda \\ &= \langle \xi_1 | \xi_2 \rangle(\theta), \end{aligned}$$

where the third equality follows from the Poisson summation formula. It is clear that Ψ has dense image since, on the suitable subspace of smooth sections of $\mathbb{R} \times_{\mathbb{Z}} \ell^2(\mathbb{Z})$ with rapid decay in the fibers, Ψ can be inverted by setting

$$(\Psi^{-1}(\eta))(\lambda) = \eta(\lambda)_0$$

($\eta \in \mathcal{E}'$, $\lambda \in \mathbb{R}$). So Ψ extends to an isometric isomorphism of Hilbert $C(S^1)$-modules. Set $\mathcal{F} = \Psi \tilde{E} \Psi^{-1}$; then

$$(\mathcal{F}\eta)(\lambda)_n = 2\pi(\lambda + n) \cdot \eta(\lambda)_n.$$

We still have to identify the unbounded Kasparov element $(\mathcal{E}', \mathcal{F}) \in KK_1(\mathbb{C}, C(S^1))$ with the generator of $K_1(C(S^1))$. For B a unital C*-algebra, Kucerovsky ([Kuc94], Chapter 6) has shown that, when $KK_1(\mathbb{C}, B)$ is described by means of unbounded elements, one may realize explicitly the isomorphism

$$KK_1(\mathbb{C}, B) \simeq K_1(B)$$

by means of the Cayley transform. If $(\mathcal{E},\mathcal{D})$ is an unbounded element of $KK_1(\mathbb{C},B)$, meaning that $\mathcal{E}$ is a Hilbert B-module and that $\mathcal{D}$ is an unbounded self-adjoint operator on $\mathcal{E}$ such that $(\mathcal{D}^2+1)^{-1}$ is compact in the sense of C*-modules, i.e. it belongs to the ideal $\mathcal{K}(\mathcal{E})$ of compact operators; then $U = \frac{i\mathcal{D}+1}{i\mathcal{D}-1}$ is a unitary operator equal to 1 modulo $\mathcal{K}(\mathcal{E})$. Hence U defines an element in $K_1(\mathcal{K}(\mathcal{E})) = K_1(B)$.

We now come back to the element $(\mathcal{E}',\mathcal{F}) \in KK_1(\mathbb{C}, C(S^1))$. The Cayley transform of $\mathcal{F}$ is the unitary operator U_1 on $\mathcal{E}'$ given by

$$\begin{aligned}(U_1\eta)(\lambda)_n &= \tfrac{2\pi i(\lambda+n)+1}{2\pi i(\lambda+n)-1}\cdot \eta(\lambda)_n\\ &= -(\exp(2i\arctan 2\pi(\lambda+n)))\cdot\eta(\lambda)_n\end{aligned}$$

$(\eta\in\mathcal{E}')$. Consider now the function

$$f_0(\lambda) = \begin{cases} -1 & if \quad \lambda \le -\frac{1}{2} \\ 2\lambda & if \quad -\frac{1}{2}\le\lambda\le\frac{1}{2} \\ 1 & if \quad \frac{1}{2}\le\lambda \end{cases}$$

and the family of unitary operators U_s $(s\in[0,1])$ on $\mathcal{E}'$ given by

$$(U_s\eta)(\lambda)_n = -(\exp(2is\arctan 2\pi(\lambda+n) + \pi i(1-s)f_0(\lambda+n)))\cdot\eta(\lambda)_n.$$

For every $s\in[0,1]$, the operator $U_s - 1$ belongs to $\mathcal{K}(\mathcal{E}')$, so this family defines a unique element in $K_1(C(S^1))$. Let us look at U_0; for $-\frac{1}{2}\le\lambda<\frac{1}{2}$, we get

$$(U_0\eta)(\lambda)_n = \begin{cases} \eta(\lambda)_n & if\, n\neq 0 \\ -exp\,2\pi i\lambda\cdot\eta(\lambda)_0 & if\, n=0. \end{cases}$$

This makes it clear that the class of U_0 in $K_1(C(S^1))$ is nothing but the canonical generator (indeed, parametrizing S^1 with $[0,1[$ instead of $[-\frac{1}{2},\frac{1}{2}[$, with $\mu=\lambda+\frac{1}{2}$ we get $-exp\,2\pi i\lambda = exp\,2\pi i\mu$). □

5. Lowest Dimensional Part of μ_1^Γ

Recall that Γ^{ab} denotes the abelianized group of Γ, and that $\kappa_\Gamma : \Gamma^{ab}\to K_1(C_r^*\Gamma)$ denotes the canonical homomorphism, induced by the map $\tilde{\kappa}_\Gamma : \Gamma \to K_1(C_r^*\Gamma)$ coming from the inclusion of Γ in the unitary group of $C_r^*\Gamma$.

In this section, we will construct a homomorphism

$$\beta_t : \Gamma^{ab} \to RK_1^\Gamma(\underline{E}\Gamma)$$

such that $\mu_1^\Gamma\circ\beta_t = \kappa_\Gamma$. This was previously done by Natsume [Nat88], under the assumption that Γ is torsion-free; this assumption was needed in order to be able to replace $RK_1^\Gamma(\underline{E}\Gamma)$ by $RK_1(B\Gamma)$.

5.1. Definition of β_t

Our definition of β_t will be in two steps: first, we define a (set-theoretic!) map $\tilde{\beta}_t : \Gamma \to RK_1^\Gamma(\underline{E}\Gamma)$; next, we prove that $\tilde{\beta}_t$ is a group homomorphism. Since the target group $RK_1^\Gamma(\underline{E}\Gamma)$ is abelian, $\tilde{\beta}_t$ factors through the desired homomorphism β_t.

To define $\tilde{\beta}_t$, we notice that every element $\gamma \in \Gamma$ defines a unique group homomorphism $\alpha_\gamma : \mathbb{Z} \to \Gamma$ by the requirement

$$\alpha_\gamma(1) = \gamma.$$

As in the proof of Proposition 4.1, let $x = (L^2(S^1), M, D)$ be the (unbounded picture of the) generator of $RK_1^{\mathbb{Z}}(\underline{E}\mathbb{Z}) = K_1(S^1) \simeq \mathbb{Z}$. By functoriality (see Theorem 1.1), we get a homomorphism $(\alpha_\gamma)_* : RK_1^{\mathbb{Z}}(\underline{E}\mathbb{Z}) \to RK_1^\Gamma(\underline{E}\Gamma)$ and we set:

$$\tilde{\beta}_t(\gamma) = (\alpha_\gamma)_*(x).$$

We begin by giving another description of this map $\tilde{\beta}_t$. Recall that $E\Gamma$ denotes the universal covering space of $B\Gamma$. Since the Γ-action on $E\Gamma$ is proper, there is a Γ-equivariant continuous map $\phi : E\Gamma \to \underline{E}\Gamma$, unique up to homotopy. Lemma 3.2 shows that, restricted to any Γ-compact subset $\tilde{X}$ of $E\Gamma$, the map ϕ is proper. By functoriality we have a map

$$\phi_* : KK_1^\Gamma(C_0(\tilde{X}), \mathbb{C}) \to RK_1^\Gamma(\underline{E}\Gamma).$$

Let X be the image of $\tilde{X}$ under the covering map $E\Gamma \to B\Gamma$. Since Γ acts freely on $\tilde{X}$, we have an identification:

$$KK_1^\Gamma(C_0(\tilde{X}), \mathbb{C}) \simeq KK_1(C_0(X), \mathbb{C}).$$

Fix $\gamma \in \Gamma$, and view γ as a loop in $B\Gamma$, i.e. as a continuous map $\gamma : S^1 \to B\Gamma$; set $X = \gamma(S^1)$. Then $\gamma_*(x) \in KK_1(C_0(X), \mathbb{C})$; applying ϕ_* we get an element in $RK_1^\Gamma(\underline{E}\Gamma)$. Thus we define:

$$\tilde{B}_t : \Gamma \to RK_1^\Gamma(\underline{E}\Gamma) : \gamma \mapsto \phi_*(\gamma_*(x)).$$

It is clear from the definition that $\tilde{B}_t$ factors through the canonical map $\phi_* : RK_1(B\Gamma) \to RK_1^\Gamma(\underline{E}\Gamma)$.

Lemma 5.1. *The maps $\tilde{\beta}_t$ and $\tilde{B}_t$ coincide. Moreover they vanish on torsion elements of Γ.*

Proof. Fix $\gamma \in \Gamma$; we distinguish 2 cases.

1) γ has infinite order in Γ. Then $\tilde{\beta}_t(\gamma) = (\alpha_\gamma)_*(x)$ is described by first considering $\tilde{X} = \Gamma \times_{\mathbb{Z}} \mathbb{R}$ with the Kasparov triple $\tilde{x}$ induced from $(L^2(\mathbb{R}), \tilde{M}, \tilde{D})$. Let $\psi : \tilde{X} \to \underline{E}\Gamma$ be a Γ-equivariant continuous map; from Lemma 3.1, it follows that

$$(\alpha_\gamma)_*(x) = \psi_*(\tilde{x}).$$

On the other hand, since Γ acts freely on $\tilde{X}$, the map ψ factors through $E\Gamma$, i.e. there exists a Γ-equivariant continuous map $\tilde{\psi} : \tilde{X} \to E\Gamma$ such that $\psi = \phi \circ \tilde{\psi}$. Then $\gamma_*(x)$ is described by $\tilde{\psi}_*(\tilde{x})$, so that by functoriality:

$$\tilde{B}_t(\gamma) = \phi_*(\gamma_*(x)) = \phi_*(\tilde{\psi}_*(\tilde{x})) = \psi_*(\tilde{x}) = (\alpha_\gamma)_*(x) = \tilde{\beta}_t(\gamma).$$

2) γ has finite order $n \geq 1$. Since α_γ factors through $\mathbb{Z}/n\mathbb{Z}$, it follows by functoriality that $(\alpha_\gamma)_* : RK_1^{\mathbb{Z}}(\underline{E}\mathbb{Z}) \to RK_1^{\Gamma}(\underline{E}\Gamma)$ factors through

$$RK_1^{\mathbb{Z}/n\mathbb{Z}}(\underline{E}\mathbb{Z}/n\mathbb{Z}) = K_1^{\mathbb{Z}/n\mathbb{Z}}(pt) = 0.$$

Hence $\tilde{\beta}_t(\gamma) = 0$.

Consider now the map $\gamma_* : RK_1^{\mathbb{Z}}(E\mathbb{Z}) \to RK_1^{\Gamma}(E\Gamma)$; it factors through a map $\tilde{\gamma}_* : RK_1^{\mathbb{Z}/n\mathbb{Z}}(E\mathbb{Z}/n\mathbb{Z}) \to RK_1^{\Gamma}(E\Gamma)$. Denote by ϕ_n the unique map from $E\mathbb{Z}/n\mathbb{Z}$ to $\underline{E}\mathbb{Z}/n\mathbb{Z} = pt$. Because of naturality, there is a commutative diagram

$$\begin{array}{ccc}
RK_1^{\mathbb{Z}/n\mathbb{Z}}(E\mathbb{Z}/n\mathbb{Z}) & \xrightarrow{\tilde{\gamma}_*} & RK_1^{\Gamma}(E\Gamma) \\
(\phi_n)_* \downarrow & & \downarrow \phi_* \\
K_1^{\mathbb{Z}/n\mathbb{Z}}(pt) & \longrightarrow & RK_1^{\Gamma}(\underline{E}\Gamma)
\end{array}$$

But again $K_1^{\mathbb{Z}/n\mathbb{Z}}(pt) = 0$, so that $\tilde{B}_t(\gamma) = \phi_*(\gamma_*(x)) = 0$. □

Proposition 5.2. *The map $\tilde{\beta}_t : \Gamma \to RK_1^{\Gamma}(\underline{E}\Gamma)$ is a group homomorphism.*

Proof. In view of the preceding lemma, we have to show that $\tilde{B}_t : \Gamma \to RK_1^{\Gamma}(\underline{E}\Gamma)$ is a group homomorphism. Fix $\gamma_1, \gamma_2 \in \Gamma$ and view γ_1, γ_2 as continuous maps $S^1 \to B\Gamma$. Set $X = \gamma_1(S^1) \cup \gamma_2(S^1)$. For $i = 1, 2$, the element $(\gamma_i)_*(x) \in K_1(X)$ is described by the triple $(L^2(S^1), \pi_i, D)$ where, for $f \in C(X)$, the operator $\pi_i(f)$ is pointwise multiplication by $f \circ \gamma_i$ on $L^2(S^1)$, and $D = \frac{1}{i} \cdot \frac{d}{d\theta}$. Similarly, $(\gamma_1\gamma_2)_*(x)$ is described by $(L^2(S^1), \pi, D)$, where $\pi(f)$ is pointwise multiplication by $f \circ \gamma_1\gamma_2$; here $\gamma_1\gamma_2 : S^1 \to X$ denotes the product loop of γ_1 and γ_2. It will be enough to show that

$$(\gamma_1\gamma_2)_*(x) = (\gamma_1)_*(x) + (\gamma_2)_*(x)$$

in $K_1(X) = KK_1(C_0(X), \mathbb{C})$. For this, consider the doubling unitary:

$$U : \left\{ \begin{array}{ccl} L^2(S^1) \oplus L^2(S^1) & \to & L^2(S^1) \\ (\xi_1, \xi_2) & \mapsto & \left(\theta \mapsto \left\{ \begin{array}{ll} \sqrt{2}\xi_1(2\theta) & if\, 0 \leq \theta \leq \frac{1}{2}; \\ \sqrt{2}\xi_2(2\theta - 1) & if\, \frac{1}{2} \leq \theta \leq 1. \end{array} \right. \right) \end{array} \right.$$

The inverse of U is given on $\xi \in L^2(S^1)$ by the formulae:

$$\left\{ \begin{array}{lcl} (U^*\xi)_1(\theta) & = & \frac{1}{\sqrt{2}}\xi(\frac{\theta}{2}) \\ (U^*\xi)_2(\theta) & = & \frac{1}{\sqrt{2}}\xi(\frac{\theta+1}{2}), \end{array} \right.$$

for $\theta \in S^1$. Since the product loop $\gamma_1\gamma_2$ is given by

$$(\gamma_1\gamma_2)(\theta) = \left\{ \begin{array}{ll} \gamma_1(2\theta) & if\, 0 \leq \theta \leq \frac{1}{2} \\ \gamma_2(2\theta - 1) & if\, \frac{1}{2} \leq \theta \leq 1, \end{array} \right.$$

one sees that $U(\pi_1 \oplus \pi_2)U^* = \pi$, and that $U(D \oplus D)U^* = \frac{D}{2}$. This shows that $(\gamma_1)_*(x) + (\gamma_2)_*(x)$ is unitarily equivalent to the triple $(L^2(S^1), \pi, \frac{D}{2})$, which in turn is trivially homotopic to $(\gamma_1\gamma_2)_*(x)$. The result follows. □

As we already mentioned, since $RK_1^\Gamma(\underline{E}\Gamma)$ is an abelian group, the homomorphism $\tilde{\beta}_t : \Gamma \to RK_1^\Gamma(\underline{E}\Gamma)$ factors through a homomorphism

$$\beta_t : \Gamma^{ab} \to RK_1^\Gamma(\underline{E}\Gamma).$$

5.2. Proof of Theorem 1.4

Since the homomorphism β_t is already constructed, it remains to prove $\kappa_\Gamma = \mu_1^\Gamma \circ \beta_t$. Clearly, it suffices to prove that, as group homomorphisms $\Gamma \to K_1(C_r^*\Gamma)$, one has

$$\tilde{\kappa}_\Gamma = \mu_1^\Gamma \circ \tilde{\beta}_t.$$

Fix $\gamma \in \Gamma$, and denote again by $\alpha_\gamma : \mathbb{Z} \to \Gamma$ the unique homomorphism such that $\alpha_\gamma(1) = \gamma$. Consider the diagram:

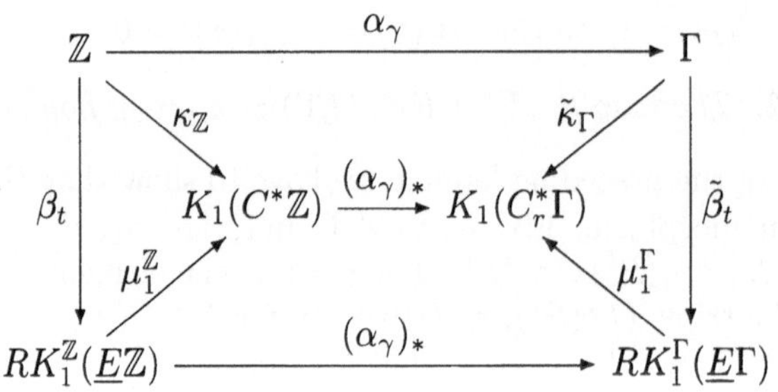

We have $(\alpha_\gamma)_* \circ \kappa_\mathbb{Z} = \tilde{\kappa}_\Gamma \circ \alpha_\gamma$ trivially, $\tilde{\beta}_t \circ \alpha_\gamma = (\alpha_\gamma)_* \circ \beta_t$ by the very definition of $\tilde{\beta}_t$, and $(\alpha_\gamma)_* \circ \mu_1^\mathbb{Z} = \mu_1^\Gamma \circ (\alpha_\gamma)_*$ by Theorem 1.1. By diagram chasing, one sees that $\tilde{\kappa}_\Gamma = \mu_1^\Gamma \circ \tilde{\beta}_t$ follows from the analogous result for $\mathbb{Z}$, i.e. $\kappa_\mathbb{Z} = \mu_1^\mathbb{Z} \circ \beta_t$. This in turn follows from Proposition 4.1 and its proof. □

Remark 5.3. 1) It follows from Theorem 1.4 that, if γ is a torsion element in Γ, then $\tilde{\kappa}_\Gamma(\gamma) = 0$. This fact can easily be proved directly, see Proposition 2 in [BV96].

2) The *strong Novikov conjecture* is the statement that

$$\mu_i^\Gamma \circ \phi_* : RK_i(B\Gamma) \simeq RK_i^\Gamma(E\Gamma) \to K_i(C_r^*\Gamma)$$

is rationally injective for $i = 0, 1$ (see [BCH94], p. 276). Since β_t factors through ϕ_*, and since κ_Γ is rationally injective (see [EN87], [BV96]), it follows from Theorem 1.4 that $\mu_1^\Gamma \circ \phi_*$ is always rationally injective on the image of $\Gamma^{ab} \simeq H_1(\Gamma, \mathbb{Z})$ in $RK_1(B\Gamma)$, i.e. on the lowest dimensional part of $\mu_1^\Gamma \circ \phi_*$.

6. Appendix: The Assembly Map in the Unbounded Picture

Dan Kucerovsky

The starting point for Kasparov KK-theory is an abstraction and axiomatization of the main properties of zeroth order elliptic operators, whereas in unbounded Kasparov theory (denoted Ψ), one uses first order operators instead.

The Kasparov product is a generalization of the "sharp product" introduced by Atiyah and Singer [AS63] in their proof of an index theorem, and this sharp product is easier to define for first order operators than for zeroth order operators. To be precise, it has been proven by Baaj and Julg [BJ83] that the Kasparov product in fact reduces to a sharp product (a graded tensor product) when written in terms of unbounded operators: their result is that

$$[D_1] \otimes [D_2] = [D_1 \otimes 1 + 1 \otimes D_2]$$

in certain special cases.

Kasparov's original approach [Kas81] to the product was to show the existence of operators M and N such that

$$[F_1] \otimes [F_2] = [M^{1/2}(F_1 \otimes 1) + N^{1/2}(1 \otimes F_2)]$$

The operators N and M have very special properties and are not easy to construct explicitly, furthermore, there are some technical complications coming from the fact that the tensor product $1 \otimes F_2$ is not well-defined in certain cases of interest, one therefore has to stabilize the Hilbert modules involved and this makes it even more difficult to explicitly determine the product cycle.

It seems plausible that M and N would be easier to construct in the unbounded picture. However, here the matter rested until the discovery of the connection approach to the Kasparov product due to Connes and Skandalis [CS84]. They found the following criterion for $(E_1 \otimes E_2, \phi_1 \otimes 1, F)$ to be the Kasparov product of (E_1, ϕ_1, F_1) and (E_2, ϕ_2, F_2) :

1. $FT_x - (-1)^{\partial x} T_x F_2$ is compact (where T_x is the tensoring operator $T_x : y \mapsto x \otimes y$).
2. $\phi_1(a)[F, F_1 \otimes 1]\phi_1(a)^*$ is positive modulo compact operators for all $a \in A$.

Given this result, it is possible to see what the counterpart in terms of unbounded cycles should be. Roughly speaking, the result of using unbounded cycles instead of bounded ones is that, quite generally, bounded operators are replaced by unbounded ones, and compactness conditions are replaced by boundedness conditions. This has the advantage that not only is boundedness often easier to prove than compactness, but the Kasparov product can be simpler to compute.

We briefly summarize both the bounded and unbounded pictures of Kasparov theory in the following table. In the table, a generic cycle in bounded KK-theory is denoted (E, ϕ, F), a Kasparov product in the bounded picture is denoted $(E_1, \phi_1, F_1) \otimes (E_2, \phi_2, F_2) = (E_1 \otimes E_2, \phi_1 \otimes 1, F)$, and the corresponding objects in the unbounded picture are denoted similarly but with D, D_1, and D_2 instead of F, F_1 and F_2.

Bounded picture: $KK(A,B)$	**Unbounded picture: $\Psi(A,B)$**
Fredholm condition:	
$\phi(a)(1-F^2)$ is compact	$\phi(a)(\lambda - D)^{-1}$ is compact for some λ.
Commutator condition:	
$[F,\phi(a)]$ is compact	$[D,\phi(a)]$ is bounded
Self-adjointness condition:	
$\phi(a)(F^* - F)$ is compact	$D^* - D$ is bounded.
Connection condition (where T_x is the tensoring operator $T_x : y \mapsto x \otimes y$.):	
$FT_x - (-1)^{\partial x} T_x F_2$ is compact.	$DT_x - (-1)^{\partial x} T_x D_2$ is bounded. It is enough to have boundedness for all x in some dense subset of $\phi_1(A)E_1$.
Positivity condition:	
$\phi(a)[F, F_1 \otimes 1]\phi(a)^*$ is positive modulo compact operators	$[D, D_1 \otimes 1]$ is bounded below. It is enough to have semiboundedness in the sense of quadratic forms on the domain of definition.
Degenerate cycles	
The representation ϕ commutes with F and $F^2 = 1$.	The representation ϕ commutes with D and D has a gap in its spectrum.
"Compact perturbation" of Kasparov cycles:	
Cycles (E,ϕ,F) and (E,ϕ,F') are equivalent if $\phi(a)(F - F')$ is compact.	The corresponding condition in the unbounded picture is that $D - D'$ is bounded on the common domain of D and D'.
Cycle homotopy:	
A cycle in $KK(A, B[0,1])$ defines a homotopy of the cycles given by evaluation at the endpoints of [0,1].	In the unbounded picture, the definition is the same, except that the cycle defining the homotopy is unbounded.
Operator homotopy:	
The special case of cycle homotopy in which the representation ϕ and Hilbert module E remain constant through the homotopy; and the operators F_0,F_1 are linked by a norm-continuous path in $\mathcal{L}(E)$.	In the unbounded picture, operator homotopies are also a special case of cycle homotopy, with the extra condition that the Cayley transform of the operator implementing the homotopy must correspond to a norm-continuous family of operators.

Equivariant cycles

Let α_g be the action of a locally compact second countable topological group, covariant with respect to the representation ϕ.

In the bounded picture, the requirement is that the function $g \mapsto (\alpha_g(F) - F)\phi(a)$ is norm-continuous and compact at every point.	In the unbounded picture, the function $g \mapsto (\alpha_g(D) - D)$ is bounded at each point and pointwise continuous in the sense that $g \mapsto (\alpha_g(D) - D)e$ is continuous for each e in the domain of D.

The equivalence relation for Kasparov cycles can be given in any of several standard forms; namely:

1. Cycle homotopy; or
2. Operator homotopy plus addition of degenerate cycles; or
3. Compact perturbation plus addition of degenerate cycles and unitary equivalence.

The equivalence of these three forms is a non-obvious but very useful result. This ends our brief outline of bounded and unbounded KK-theory. For a more precise description, we refer the reader to [Bla86], [Kas81], [Kas95] for the bounded theory, and [BJ83], [Kuc94], [Kuc97, Kuc] for the unbounded theory.

We now discuss Chapter 2 of Valette's Notes from the point of view of unbounded KK-theory. The main point of that chapter is to show that a certain map from a direct limit of equivariant K-homology groups to equivariant K-theory is well-defined. It is sufficient to define a map

$$\tilde{\mu}^\Gamma : K_i^\Gamma(X) \longrightarrow K_i(C^*\Gamma)$$

for Γ-compact subsets X of $\underline{E}\Gamma$, provided that this map commutes with the inclusion maps used to define the direct limit $RK_i^\Gamma(\underline{E}\Gamma)$. In terms of unbounded Kasparov theory, the explicit definition of the map $\tilde{\mu}^\Gamma$ is:

$$\begin{array}{rcl} \tilde{\mu}^\Gamma : \Psi_\Gamma(C_0(X), \mathbb{C}) & \longrightarrow & \Psi_\Gamma(\mathbb{C}, C^*\Gamma) \\ (\mathcal{H}, \pi, D) & \mapsto & (\mathcal{E}, D) \end{array}$$

where $\mathcal{E}$ is the $\mathbb{Z}_2$-graded Hilbert $C^*\Gamma$ module obtained by completing $\pi(C_c(X))\mathcal{H}$ with respect to the $C^*\Gamma$-valued inner product $\langle \xi_1, \xi_2 \rangle (\gamma) := \langle \xi_1, \gamma\xi_2 \rangle$ as in section 2.3. It is not obvious that this map $\tilde{\mu}^\Gamma$ is well-defined, and the easiest way to see that it does map a cycle to a cycle is by an indirect approach.

We define a large semigroup $U_\Gamma(A, B)$ that contains $\Psi_\Gamma(A, B)$:

Definition 6.1. The semigroup $U_\Gamma(A,B)$ is given by triples $(E \oplus E, \phi, \begin{pmatrix} 0 & D \\ D & 0 \end{pmatrix})$ with the direct sum operation, where E is any Hilbert B-module with an action of Γ, ϕ is an (equivariant) representation of A on E, and D is a properly supported equivariant regular operator[5] on E with $[D, \phi(a)]$ bounded for all a in some dense subset of A.

We don't need the equivariance condition, assuming the boundedness of $[D, \gamma]$ would be enough, but this additional condition, which is also assumed in the main part of this paper, makes the calculations more clear by suppressing some terms involving $[D, \gamma]$ that would otherwise arise. The reason for introducing this semigroup is only that it is convenient to have a large semigroup for which some form of the Kasparov equivalence relation makes sense. In fact, we could prove our main result without any reference to this semigroup, which is only used as a convenient setting for proving compactness of the resolvent of a certain operator.

Definition 6.2. The equivalence relation *ubd* on $U(A,B)$ is generated by unitary equivalence, perturbation by bounded operators, and addition of degenerate Kasparov cycles.

Then we see that the Kasparov group is stable under equivalence within $U_\Gamma(A,B)$:

Proposition 6.3. *$U_\Gamma(A,B)/ubd$ contains $\Psi_\Gamma(A,B)$.*

This proposition is an equivariant version of a lemma in [Kuc].

Proof. In terms of cycles, this proposition says that a triple $(E \oplus E, \phi, \begin{pmatrix} 0 & D \\ D & 0 \end{pmatrix})$ is an unbounded cycle if and only if it is equivalent to an unbounded cycle. This is obvious except for the compactness of the resolvent, which follows from the facts that unitary equivalence preserves the compact operators on a Hilbert module, and that $\phi(a)(\lambda - T)^{-1}$ is compact for some λ if and only if this is true for $\phi(a)(\lambda' - T + B)^{-1}$, where B is some bounded operator. □

The map $\tilde{\mu}^\Gamma$ is certainly at least a map into $U_\Gamma(\mathbb{C}, C^*\Gamma)$, and we will use a unitary equivalence and bounded perturbation to show that the image is in $\Psi^\Gamma(\mathbb{C}, C^*\Gamma)$ after equivalence.

[5] The definition of regularity is that an operator is regular if and only if its graph is orthogonally complemented. It follows that a bounded operator is regular if and only if it is adjointable. The reason for preferring regularity to adjointability is that regularity is defined in terms of the graph of the operator, and hence the definition can be immediately generalized to the case of unbounded operators. All Hilbert space operators are regular, but there are usually many non-regular operators on a Hilbert module.

We remind the reader that $\tilde{\mathcal{H}}$ is the $C^*\Gamma$-module completion of $C_c(\Gamma, \mathcal{H})$ with respect to the $\mathbb{C}\Gamma$-valued inner product $\langle \xi, \mu \rangle (\sigma) := \sum_{\gamma \in \Gamma} \langle \xi(\gamma), \mu(\gamma\sigma) \rangle$, and that in Lemmas 2.13 and 2.14 we defined another inner product $C_c(X) \times C_c(X) \longrightarrow C_c(X \times \Gamma)$ and then constructed a projection $p \in C_c(X \times \Gamma)$ by defining $p := \langle h, h \rangle$ where $h \in C_c(X)$ is a function with $\sum_{\gamma \in \Gamma} h(\gamma x)^2 = 1$ for all x. We now check that the function h can be chosen to have bounded commutator with a given unbounded operator that comes from a triple in $U_\Gamma(\mathbb{C}, C^*\Gamma)$.

Recalling that the function h^2 can be constructed by choosing any positive f which is nonzero on a compact global slice [Pal61] (which exists since X is a Γ-compact Γ-space with proper action of Γ) and then performing an averaging operation that will be described; we can also assume that given an unbounded selfadjoint operator D that comes from a triple, the commutator $[f, D]$ is bounded on the domain of D. The function h^2 is defined by

$$h^2(x) := \frac{f(x)^2}{\sum_{\gamma \in \Gamma} f(\gamma x)^2}.$$

Since D is properly supported, there is a g such that $g[f^2, D] = [f^2, D]$, and taking adjoints shows that $g[f^2, D]g = [f^2, D]$. In particular, the function $f_\Gamma^2 := \sum_{\gamma \in \Gamma} \gamma f^2 \gamma^{-1}$ has bounded commutator with D, since $A_g([f^2, D]) = [f_\Gamma^2, D]$ is bounded. Therefore

$$\pm i[1/f_\Gamma, D] \leq \|[f_\Gamma^2, D]\| f_\Gamma^{-3}$$

is bounded, by a form of Powers' identity[Pow75, Kuc], and hence $[h, D] = [f/f_\Gamma, D]$ is bounded.

Lemma 6.4. *There is a unitary $\beta : \tilde{\pi}(p)\tilde{\mathcal{H}} \longrightarrow \mathcal{E}$ which almost commutes with operators coming from triples $(\mathcal{E}, D)$ in $U_\Gamma(\mathbb{C}, C^*\Gamma)$, in the sense that $\beta\tilde{\pi}(p)D - D\beta : \tilde{\pi}(p)\tilde{\mathcal{H}} \longrightarrow \mathcal{E}$ is bounded.*

Proof. The map β is defined in the proof of Lemma 2.15, where it is also shown that $\langle \beta x, \beta x \rangle_{\mathcal{E}} = \langle x, x \rangle_{\tilde{\mathcal{H}}}$, which implies by the polarization identity that β is an unitary.

For the second part of the lemma, we recall that $[\pi(h), D]$ can be assumed bounded on the domain of D, and since β is unitary it is enough to show that $\beta\tilde{\pi}(p)D\beta^* - D : \mathcal{E} \longrightarrow \mathcal{E}$ is bounded. Suppressing the representation $\tilde{\pi}$, we write

$$\begin{aligned} \beta p D \beta^* &= \textstyle\sum_{\gamma \in \Gamma} \gamma h D h \gamma^{-1} \\ &= A_g([h, D]h) + \textstyle\sum_{\gamma \in \Gamma} \gamma D h^2 \gamma^{-1} \quad , \\ &= A_g([h, D]h) + D \end{aligned}$$

where we have first used the proper support of D to find a $g \in C_0(X)$ such that $hg = h$ and $gDh = Dh$, and then we have used the equivariance of D and the averaging property of h to show that $\sum_{\gamma \in \Gamma} \gamma D h^2 \gamma^{-1} = D$. Since $A_g([h, D]h)$ is bounded, we are done. □

Remark 6.5. Since $A_h(T) = \beta T \beta^*$ for $T : \mathcal{H} \longrightarrow \mathcal{H}$, we see that the "averaging operator" A_f is actually a unitary transformation in the special case $f = h$, and therefore can be applied without losing information. Furthermore, A_h commutes with the functional calculus.

We can now give a quick proof that an element $(\mathcal{E}, D)$ of the image of $\tilde{\mu}^\Gamma$ is a cycle. First of all, if we consider a map from $\Psi_\Gamma(C_0(X), \mathbb{C})$ to $\Psi_\Gamma(\mathbb{C}, C^*\Gamma)$ given by $(\mathcal{H}, \pi, D) \mapsto (\tilde{\mathcal{H}}, \tilde{\pi}(p) : \mathbb{C} \longrightarrow \mathcal{L}, D)$, then it is quite clear that an element of the image is in fact a cycle. The main question is whether or not the resolvent $R := \tilde{\pi}(p)(i + D)^{-1}$ is compact on $\tilde{\mathcal{H}}$, but regarding $\tilde{\mathcal{H}}$ as a function space over Γ, we see that R is compact at every point, and is compactly supported because

$$\tilde{\pi}(p)\xi := \pi(x \mapsto \overline{h(x)} h(\gamma^{-1}x))\xi(\gamma), \quad \xi \in C_c(\Gamma, \mathcal{H})$$

where h is compactly supported and Γ acts properly.

Now we use the preceeding lemma to give the equivalences

$$\begin{aligned} (\tilde{\mathcal{H}}, \tilde{\pi}(p) : \mathbb{C} \longrightarrow \mathcal{L}, D) &\sim_d (\tilde{\mathcal{H}}, \tilde{\pi}(p)D) \\ &\sim_u (\mathcal{E}, \beta\tilde{\pi}(p)D\beta^*) \\ &\sim_b (\mathcal{E}, D) \end{aligned}$$

proving that the map from $\Psi_\Gamma(C_0(X), \mathbb{C})$ to $\Psi_\Gamma(\mathbb{C}, C^*\Gamma)$ that we just gave is in fact the Baum–Connes map $\tilde{\mu}^\Gamma$.

Dan Kucerovsky
Dep. of Mathematics and Statistics
University of New Brunswick
Fredericton, New Brunswick
CANADA E3B 5A3
dan@math.unb.ca

References

[AD87] C. Anantharaman-Delaroche. Systèmes dynamiques non commutatifs et moyennabilité. *Math. Ann.* **279** (1987), 297–315.

[Arv76] W. Arveson. *An invitation to C*-algebra.* Springer, 1976.

[AS63] M.F. Atiyah and I.M. Singer. The index of elliptic operators on compact manifolds. *Bull. Amer. Math. Soc.* **69** (1963), 422–433.

[BBV99] C. Béguin, H. Bettaieb, and A. Valette. K-theory for the C*-algebras of one-relator groups. *K-theory* **16** (1999), 277–298.

[BC88a] P. Baum and A. Connes. Chern character for discrete groups. In *A fête of topology (Academic Press, pp. 163–232)*, 1988.

[BC88b] P. Baum and A. Connes. K-theory for discrete groups. In *Operator algebras and application* (London Math. Soc. lecture notes ser. **135**, 1–20), 1988.

[BC00] P. Baum and A. Connes. Geometric K-theory for Lie groups and foliations. *Enseign. Math.* **46** (2000), 3–42. First distributed: 1982.

[BCdlH94] M.E.B. Bekka, M. Cowling, and P. de la Harpe. Some groups whose reduced C*-algebra is simple. *Publ. Math. I.H.E.S.*, **80** (1994), 117–134.

[BCH94] P. Baum, A. Connes, and N. Higson. Classifying spaces for proper actions and K-theory of group C*-algebras. In *C*-algebras 1943–1993: a fifty year celebration* (Contemporary Mathematics **167**, 241–291), 1994.

[BJ83] S. Baaj and P. Julg. Théorie bivariante de Kasparov et opérateurs non bornés dans les C*-modules hilbertiens. *C.R.Acad.Sci.Paris* **296** (1983), 875–878.

[Bla86] B. Blackadar. *K-theory for operator algebras.* MSRI publications 5, Springer-Verlag, 1986.

[BV96] H. Bettaieb and A. Valette. Sur le groupe K_1 des C*-algèbres réduites de groupes discrets. *C.R. Acad. Sci. Paris*, **322** (1996), 925–928.

[Con94] A. Connes. *Noncommutative geometry.* Academic Press, 1994.

[CS84] A. Connes and G. Skandalis. The longitudinal index theorem for foliations. *Pub. Res. Inst. Math. Sci. Kyoto Univ.* **20** (1984), 1139–1183.

[Cun83] J. Cuntz. K-theoretic amenability for discrete groups. *J. für reine angew. Math.* **344** (1983), 180–195.

[Dix77] J. Dixmier. *C*-algebras.* North Holland, 1977.

[DL98] J.F. Davis and W. Lück. Spaces over a category and assembly maps in isomorphism conjectures in K- and L-theory. *K-theory* **15** (1998), 201–252.

[EN87] G. Elliott and T. Natsume. A Bott periodicity map for crossed products of C*-algebras by discrete groups. *K-theory* **1** (1987), 423–435.

[FRR95] S.C. Ferry, A. Ranicki, and J. Rosenberg. A history and survey of the Novikov conjecture. In *Novikov conjectures, index theorems and rigidity* (London Math. Soc. lecture notes ser. **226**, pp. 7–66), 1995.

[Gre77] P. Green. C*-algebras of transformation groups with smooth orbit space. *Pac. J. of Math.* **72** (1977), 71–97.

[Gro93] M. Gromov. Asymptotic invariants of infinite groups. In *Geometric group theory* (G.A. Niblo and M.A. Roller, eds.), London Math. Soc. lect. notes **182**, Cambridge Univ. Press, 1993.

[HLS02] N. Higson, V. Lafforgue, and G. Skandalis. Counterexamples to the Baum–Connes conjecture. *Geom. funct. anal.* **12** (2002), 330–354.

[HP] I. Hambleton and E.K. Pedersen. Identifying assembly maps in K- and L-theory. Preprint, june 2001.

[HR00a] N. Higson and J. Roe. Amenable actions and the Novikov conjecture. *J. Reine Angew. Math. 519, 143–153* **519** (2000), 143–153.

[HR00b] N. Higson and J. Roe. *Analytic K-homology.* Oxford mathematical monographs, 2000.

[JT91] K. Knudsen Jensen and K. Thomsen. *Elements of KK-theory.* Birkhäuser, 1991.

[Jul98] P. Julg. Travaux de Higson et Kasparov sur la conjecture de Baum–Connes. In *Séminaire Bourbaki*, Exposé **841**, 1998.

[Kas75] G.G. Kasparov. Topological invariants of elliptic operators.i.K-homology. *Math. USSR Izvestija* **9** (1975), 751–792.

[Kas81] G.G. Kasparov. The operator K-functor and extensions of C*-algebras. *Math. of the USSR-Izvestija* **16** (1981), 513–572.

[Kas83] G.G. Kasparov. The index of invariant elliptic operators, K-theory, and Lie group representations. *Dokl.Akad. Nauk. USSR* **268** (1983), 533–537.

[Kas88] G.G. Kasparov. Equivariant KK-theory and the Novikov conjecture. *Invent. Math.* **91** (1988), 147–201.

[Kas95] G.G. Kasparov. K-theory, group C*-algebras, and higher signatures (Conspectus, first distributed 1981). In *Novikov conjectures, index theorems and rigidity* (London Math. Soc. lecture notes ser. **226**, 101–146), 1995.

[Kuc] D. Kucerovsky. Making Kasparov products unbounded. Preprint, 1997.

[Kuc94] D. Kucerovsky. *Kasparov products in KK-theory, and unbounded operators with applications to index theory.* PhD thesis, Magdalen College, Oxford, 1994.

[Kuc97] D. Kucerovsky. The KK-product of unbounded modules. *K-theory* **11** (1997), 17–34.

[Luc] W. Lück. On the functoriality of the source of the Baum–Connes map. Unpublished, December 1996.

[LÖ2] W. Lück. The relation between the Baum–Connes conjecture and the trace conjecture. *Inventiones. Math.* **149** (2002), 123–152.

[Laf98] V. Lafforgue. Une démonstration de la conjecture de Baum–Connes pour les groupes réductifs sur un corps p-adiques et pour certains groupes discrets possédant la propriété (t). *C.R. Acad. Sci. Paris* **327** (1998), 439–444.

[Lan95] E.C. Lance. *Hilbert C*-modules, a toolkit for operator algebraists.* London Math. Soc. lecture notes ser. **210**, 1995.

[Mat] M. Matthey. Low dimensional homology, the Baum–Connes assembly map, delocalization and the Chern character. Preprint, Muenster 2001.

[Mat00] M. Matthey. *K-theories, C*-algebras and assembly maps.* PhD thesis, Université de Neuchâtel, 2000.

[Mis] G. Mislin. Equivariant K-homology of the classifying space for proper actions. Proceedings of the Euro summer school on proper group actions, Barcelona, september 2001.

[Nat88] T. Natsume. The Baum–Connes conjecture, the commutator theorem, and Rieffel projections. *C.R. Math. Rep. Acad. Sci. Canada* **X** (1988), 13–18.

[Pal61] R. Palais. On the existence of slices for actions of non-compact Lie groups. *Ann. of Math.* **73** (1961), 295–323.

[Pas85] D.S. Passman. *The algebraic structure of group rings.* Krieger Publishing Company, 1985.

[Ped79] G.K. Pedersen. *C*-algebras and their automorphism groups.* Academic Press, 1979.

[Pie00] F. Pierrot. *K-théorie de C*-algèbres pleines de groupes de Lie et formule de multiplicité de Langlands.* PhD thesis, Paris VII, 2000.

[Pow75] R.T. Powers. A remark on the domain of an unbounded derivation of a C*-algebra. *J. Funct. Anal.* **18** (1975), 85–95.

[Rie74] M.A. Rieffel. Induced representations of C*-algebras. *Adv. in Math.* **13** (1974), 176–257.

[Rie82] M.A. Rieffel. Applications of strong Morita equivalence to transformation group C*- algebras. In *Operator algebras and applications*, Proc. Symp. Pure Math. **38**, Part 1, 299–310, 1982.

[Roe] J. Roe. Comparing analytic assembly maps. Preprint, April 2001.

[Roe96] J. Roe. *Index theory, coarse geometry, and topology of manifolds.* CBMS regional conf. ser. in Math. **90**, 1996.

[Ros83] J. Rosenberg. C*-algebras, positive scalar curvature, and the Novikov conjecture. *Publ. Math. I.H.E.S.* **58** (1983), 197–212.

[Ros84] J. Rosenberg. Group C*-algebras and topological invariants. In *Operator algebras and group representations*, vol. II. Pitman, pp. 95–115, 1984.

[Ska99] G. Skandalis. Progrès récents sur la conjecture de Baum–Connes. Contribution de Vincent Lafforgue. In *Séminaire Bourbaki*, Exposé **869**, 1999.

[Tay75] J.L. Taylor. Banach algebras and topology. In *Algebras in analysis*, pp. 118–186. Academic Press, 1975.

[Val89] A. Valette. The conjecture of idempotents: a survey of the C*-algebraic approach. *Bull. Soc. Math. Belg.*, **XLI** (1989), 485–521.

[Val98] A. Valette. On Godement's characterization of amenability. *Bull. Austral. Math. Soc.* **57** (1998), 153–158.

[Val02] A. Valette. *Introduction to the Baum–Connes conjecture.* ETHZ lectures in math., Birkhäuser, 2002.

Index